Green Energy and Technology

I. Pilatowsky · R.J. Romero · C.A. Isaza
S.A. Gamboa · P.J. Sebastian · W. Rivera

Cogeneration Fuel Cell-Sorption Air Conditioning Systems

 Springer

I. Pilatowsky, Dr.
S.A. Gamboa, Dr.
P.J. Sebastian, Dr.
W. Rivera, Dr.

Universidad Nacional Autónoma
de México
Centro de Investigación en Energía
Cerrada Xochicalco s/n Colonia Centro
62580 Temixco
Morelos
México

ipf@cie.unam.mx
sags@cie.unam.mx
sjp@cie.unam.mx
wrgf@cie.unam.mx

www.cie.unam.mx

R.J. Romero, Dr.
Universidad Autónoma del Estado
de Morelos
Centro de Investigación en Ingeniería
y Ciencias Aplicadas
Avenida Universidad 1001
62210 Cuernavaca
Morelos
México
rosenberg@uaem.mx
www.uaem.mx

C.A. Isaza, Dr.
Universidad Pontificia Bolivariana
Instituto de Energía, Materiales y Medio
Ambiente
Grupo de Energía y Termodinámica
Circular 1, no.73-34
70-01, Medellín
Colombia
cesar.isaza@upb.edu.co
www.upb.co

ISSN 1865-3529
ISBN 978-1-4471-2632-4
DOI 10.1007/978-1-84996-028-1
Springer London Dordrecht Heidelberg New York

ISSN 1865-3537 (eBook)
ISBN 978-1-84996-028-1 (eBook)

British Library Cataloguing in Publication Data
A catalogue record for this book is available from the British Library

Cover design: eStudioCalamar, Girona/Berlin

Printed on acid-free paper

Springer is part of Springer Science+Business Media (www.springer.com)

Preface

The global energy demand increases every day with increase in population and modernization of the way of life. The intense economic activity around the world depends largely on fossil fuel based primary energy. The indiscriminate use of fossil fuel based energy has inflicted severe damage to air quality, caused water contamination, and environmental pollution in general.

The exploitation of renewable energy sources has been proposed as a solution to encounter the above mentioned global problems. The major problems associated with the exploitation of renewable energy sources are their intermittency, high cost of energy conversion and storage, and low efficiency. In addition, the wide spread utilization of renewable energy leads to the culture of energy saving and rational end use.

Hybrid systems based on different renewable energy sources are becoming more relevant due to the intermittency of single primary energy sources, the increase in the final efficiency in energy conversion in a hybrid system, and the final cost reduction. Moreover, hybrid systems can satisfy the energy demand of a specific application un-interruptedly. There are different types and combinations of hybrid energy systems presently employed around the world. To mention a few, there are photovoltaic-wind energy systems, photovoltaic-thermal energy systems, wind-hydrogen-fuel cell systems, *etc*.

Combined heat and power (CHP) systems have been known for quite some time as a part of hybrid systems. The advantage of this kind of system is its high efficiency, low cost compared to other hybrid systems, and low economic impact without sacrificing continuous energy supply to the load.

This book deals with a new concept in CHP systems where a fuel cell is used for generating electricity and the heat released during the operation of the cell is used for air conditioning needs. For the CHP system considered in this book, we have chosen heat proton exchange membrane fuel cell in particular due to the temperature of the ejected and the air-conditioning needs of the CHP system.

For the authors to have a general understanding of the topic we have treated the energy and co-generation processes in detail. The thermodynamic principles governing energy conversion in general and fuel cells in particular have been treated briefly. The principles of CHP systems have been explained in detail with particular emphasis on sorption air-conditioning systems.

The authors would like to thank María Angelica Prieto and María del Carmén Huerta for their collaboration in the English grammatical review and formatting, respectively. Also the authors would like to thank Geydy Gutiérrez Urueta for her contribution in the revision and suggestions in the present work.

Mexico/Colombia   *I. Pilatowsky*
December 2010   *R.J. Romero*
                                                                                           C.A. Isaza
                                                                                       S.A. Gamboa
                                                                                     P.J. Sebastian
                                                                                          W. Rivera

Contents

Notation

A	Annual cost or payments (€ years^{-1})
A_{CF}	Annual cash flow (€ years^{-1})
A_{CI}	Annual cash income (€ years^{-1})
A_{CTR}	Annual cooling time required (€ years^{-1})
A_{DCF}	Annual discount cash flow (€ years^{-1})
A_E	Annual cost of electricity (€ years^{-1})
A_G	Annual cost of gas (€ years^{-1})
A_{NCI}	Net annual cash income (€ years^{-1})
A_S	Annual sales (€ years^{-1})
A_T	Annual amount of tax (€ years^{-1})
A_{TC}	Annual total cost (€ years^{-1})
A_{TE}	Annual total expenses (€ years^{-1})
AD	Average annual amount of depreciation (€ years^{-1})
AEP	Annual electricity cost (€ years^{-1})
AFC	Alkaline fuel cell
AFC_{H2}	Annual fuel cost of hydrogen (€ years^{-1})
ARC	Absorption refrigerating cycle
ARS	Absorption refrigerating machine
CFC	Chlorofluorocarbons
CHP	Cooling, heating, and power or combined heat and power system
C_E	Electricity unit cost (€)
C_{FC}	Fixed capital costs (€)
C_{FCR}	Fuel cell replacement cost (€)
C_G	Gas unit cost (€)
C_{H2}	Hydrogen cost (€)
C_L	Land cost (€)
C_P	Heat capacity (kJ kg^{-1} $^{\circ}$C^{-1})
C_{TC}	Total capital costs (€)
C_{WC}	Working capital costs (€)
CFR	Capital recovery factor (dimensionless)

COP	Coefficient of performance (dimensionless)
COP_{AC}	Air-conditioning COP (dimensionless)
COP_R	Coefficient of performance of refrigeration cycle (dimensionless)
COP_{T1}	Coefficient of performance for type 1 system (dimensionless)
COP_{T2}	Coefficient of performance for type 2 system (dimensionless)
DAHX	Desorber/absorber/heat exchanger
DMFC	Direct methanol fuel cell
E_t	Thermo-neutral potential (V)
E_0	Ideal standard potential (V)
E	Ideal equilibrium potential or reversible potential (V)
EOC	Efficiency of cogeneration (dimensionless)
EOT	Efficiency of tri-generation (dimensionless)
ETP	Equivalent tons of petroleum
EU	European Union
EUAC	Equivalent uniform annual cost ($€$ years^{-1})
E_{V4}	Expansion valve
F	Future worth ($€$), Faraday's constant (9.65×10^4 C mol^{-1})
f_{AF}	Annuity future worth factor (dimensionless)
f_{AP}	Annuity present worth factor (dimensionless)
f_i	Compound interest factor (dimensionless)
f_d	Discount factor (dimensionless)
FC	Fuel cell
GAX	Generator/absorber/heat exchanger
ΔG	Free energy change (kJmol^{-1})
ΔG^0	Free energy change at standard conditions (kJmol^{-1})
GDP	Gross domestic product (dimensionless)
HE	Heat exchanger
ΔH	Enthalpy change or standard enthalpy of formation (kJkg^{-1})
HRSG	Heat recovery steam generators
I_0	Investment at the beginning of the project ($€$)
$I_{0,U}$	Initial cost *per* unit or energy ($€$)
IRR	Internal rate of return (fraction or %)
i	Interest or discount rate (fraction of %) or current density
i_E	Electricity inflation (fraction or %)
i_G	Gas inflation (fraction of %)
i_L	Limiting current density (Am^{-2})
i_0	Exchange current density (Am^{-2})
K	Constant of electrical power for fuel cell (W)
m_w	Cooler flow into the FC (kgs^{-1})
MCFC	Molten carbonate fuel cell
MEA	Membrane electrode assembly
n	Number of interest periods or number of electrons (dimensionless)
NPC	Net present cost ($€$)
NPV	Net present value ($€$)
P	Present worth ($€$) or pressure (Pa)

P_e	Electrical power (W)
P_{H2}	Energy from hydrogen (kJ)
PAFC	Phosphoric acid fuel cell
PBP	Payback period (years)
PC	Compressor power capacity (kW)
PEM	Proton exchange membrane
PEMFC	Proton exchange membrane fuel cell
Q	Heat (kJ)
Q_G	Thermal energy supplied (kJ)
Q_T	Thermal energy (kJ)
Q_W	Waste heat (kJ)
ΣQ_{IN}	Energy input (kJ)
ΣQ_U	Useful energy (kJ)
R	Gas constant (8.34 J mol^{-1}K^{-1})
R_c	Cell resistance (Ohm)
S	Scrap value (€) or entropy (kJkg^{-1}°C^{-1})
S_T	Thermal source (dimensionless)
ΔS	Entropy change (kJkg^{-1}°C^{-1})
SOFC	Solid oxide fuel cell
SRC	Sorption refrigeration cycle
T	Temperature (K, °C)
T_O	Operating time (h years^{-1})
T_f	Final temperature in the fuel cell outlet (°C)
$T_{GE,out}$	Exit temperature of Absorption heat pump to fuel cell (°C)
T_i	Initial temperature at the fuel cell inlet (°C)
TDARS	Thermal driven adsorption refrigeration system
ΔT_c	Coupling temperature difference (°C)
ΔV	Volume change (m^3)
W_e	Maximum electrical work (kJ)
$W_{elect,max}$	Maximum electrical work (kJ)

Subscripts

A, AB	Absorber
C, CO	Condenser
CF	Cooling fluid
D	Desorber
E, EV	Evaporator
G, GE	Generator
HF	Heating fluid
MC	Mechanical compressor
P	Pump
R	Rectifier

RC   Recipient of condensate
RF   Fluid cooling
S   Sorber
SF   Sorber fluid
SO   Sorption
TM   Thermal machine
X   Heat exchanger

Greek Letters

α Charge transfer coefficient (dimensionless)
η Efficiency (dimensionless)
η_{act} Activation polarization (V)
η_{con} Concentration polarization (V)
η_e Thermal efficiency (dimensionless)
η_{elect} Efficiency of conversion (dimensionless)
η_{ohm} Ohmic polarization (V)
η_{In} Theoretical thermal efficiency for internal regime (dimensionless)
η_{Ex} Theoretical thermal efficiency for external regime (dimensionless)

Chapter 1
Energy and Cogeneration

1.1 Introduction

1.1.1 Energy Concept

The word energy is derived from the Greek *in* (in) and *ergon* (work). The accepted scientific energy concept has been used to reveal the common characteristics in diverse processes where a particular type of work is produced. At the most basic level, the diversity in energy forms can be limited to four: kinetics, gravitational, electric, and nuclear.

Energy is susceptible to being transformed from one form to another, where the total quantity of energy remains unchanged; it is known that: "Energy can neither be created nor destroyed, only transformed". This principle is known as the first law of thermodynamics, which establishes an energy balance in the different transformation processes.

When the energy changes from one form to another, the energy obtained at the end of the process will never be larger than the energy used at the beginning, there will always be a defined quantity of energy that could not be transformed.

The relationship of useful energy with energy required for a specific transformation is known as conversion efficiency, expressed in percent. This gives origin to the second law of the thermodynamics, which postulates that the generation of work requires a thermodynamic potential (temperature, pressure, electric charges, *etc.*) between two energy sources, where energy flows from the highest potential to the lowest, in which process there is a certain amount of energy that is not available for recovery. In general, the second law establishes the maximum quantity of energy possible that one can obtain in a transformation process, through the concept of exergetic efficiency.

The energy is the motor of humanity's social, economic, and technological development, and it has been the base for the different stages of development of society: (1) the primitive society whose energy was based on its own human energy and on the consumption of gathered foods, (2) the society of hunters, which had a nomadic character, based on the use of the combustion of the wood, (3) the primitive agricultural society, which consumed wood and used animal traction, (4) the advanced agricultural society, which consumed wood, energy derived from water and wind, and some coal and animal traction, (5) the industrial society, which consumed coal (for vapor production), wood, and some petroleum, and finally (6) the technological society, which consumes petroleum (especially for machines of internal combustion), coal, gas, and nuclear energy.

Current society depends for the most part on the energy resources derived from petroleum and due to its character of finite, high costs and problems of contamination, the energy resources should be diversified, with priority toward the renewable ones, depending on the characteristic of each country and region.

The energy at the present time is intimately related to aspects such as saving and efficient use, economy, economic and social development, and the environment, which should be analyzed in order to establish an appropriate energy politics to assure the energy supply and therefore the necessary economic growth.

1.1.2   Energy and Its Impacts

1.1.2.1   Energy and Development

In the relationship between energy and development, the consumptions of primary energy in the world, and regions, the tendencies, as well as external trade and prices are analyzed. Concerning the perspectives, the energy reserves, the modifications in the production structure, and the modifications in consumption and prices are studied.

An important aspect to consider is the analysis of the relationship between energy and development through the relationship between the energy consumption and the gross domestic product (GDP) of a country. On the one hand, the GDP is representative of the level of economic life, and on the other, it is indicative of the level of the population's life and therefore of the degree of personal well-being reached. The economic activity and the well-being imply energy consumption; in the first case, the energy would be an intermediate goods of consumption that is used in the productive processes in order to obtain goods and services, in the second case, a final goods of consumption for the satisfaction of personal necessities, such as cooking and conservation of food, illumination, transport, air conditioning, *etc.*

For a particular country, the relationship that exists among the total consumption of primary energy *per* year in a given moment, generally evaluated in equivalent tons of petroleum (ETP) and the GDP, evaluated in constant currency, gives an idea of the role of energy in the economic activity. The relationships are called energy intensity of the GDP, energy content of the GDP, or energy coefficient. However, it is verified that such a quotient is at the same time very variable, so much in the time for a country in particular, as in the space in a given moment if several countries are considered – even if they have comparable levels of economic development. This is not only completed at macroeconomic level but also for a sector or specific economic branch.

In general, both variations of the energy intensity are strongly determined by two types of factors: (1) factors that concern the national economic structure, as the nature or percentage of participation of the economic activities that compose the GDP; this is because the energy consumption for units of product is very diverse depending on the sector of the economy (agriculture, industry, transport, services, *etc.*); and (2) technological factors that refer to the type of energy technology consumed and the form used by each industry, economic sector, or consumer.

The participation of the energy sector in the GDP is usually low, although in some countries strong petroleum producers for export will spread to become bigger, with consideration of the following aspects: (1) energy availability is a necessary condition, although not sufficient for the development of economic activity and the population's well-being. (2) The energy sector is, at least potentially, one of the motors of the industrial and technological development of the country, since it is the most important national plaintiff of capital goods, inputs, and services. (3) Import requirements, and therefore of foreign currencies, are a function of the degree of integration with the productive sector. (4) To be a capital-intensive sector, it competes strongly with others in the assignment of resources. (5) Moreover, since energy projects have a very long period of maturation, equipment not used to its capacity produces restrictions and important costs in the economic activity, while an oversupply would mean a substantial deviation of unproductive funds that could be used by another sector. (6) The increasing and increasingly important participation of the energy sector in the government's fiscal revenues.

1.1.2.2 Energy and Economy

The energy analysis should not only consider the technical and political aspects, but also the economic one. From an economic point of view, energy satisfies the necessities of the final consumers and of the productive apparatus, which includes the energy sector. To arrive to the consumer it is necessary to continue by a chain of economic processes of production and distribution. These processes and the companies that carry out and administer them, configure the energy sector, being key in the economy, because it is a sector with a strong added value, very intensive in capital and technology, with an important weight in external trade, both in the producing countries as consumers of energy and in public finances.

For these characteristics the energy sector is susceptible to exercising multiple macroeconomic effects, through its investments, of the employment and added value that it generates, of the taxes that it pays or it makes people pay for its products, which produces numerous inter-industrial effects, since energy is an intermediate consumption of all branches of economic activity.

The economic analysis has three basic focuses: microeconomic, macroeconomic, and international relationships, with particular theoretical and methodological principles considering the specificities of the energy sector. The microeconomic analysis is based on economic calculation in the energy sector and administration of public energy companies. This includes (1) prices and costs of energy resources, (2) problems of the theory of the value, (3) internal prices, tarification systems and policies, (4) analysis of energy consumption and its determinants, (5) public companies and problems of energy investments, (6) evaluation of the energy projects, and (7) some alternative lines of analysis.

In macroeconomic analysis energy is considered in relation to the economic growth problem, through the perspective of planning. The central topics are: (1) energy and factors and global production functions, (2) energy and economic growth; analysis instruments and main evolutions, (3) macroeconomic implications of the evolution of prices, for example of petroleum, the question of the surplus and its use, and (4) macroeconomics, energy modeling, and planning.

In the case of international relationships, the mains topics for the study of main phenomena and energy processes are: (1) the main actors of energy scene, (2) the nature of markets and energy industries and their recent restructuring, and (3) the determination of international energy prices.

The insurance of the best selection of investment projects is an important aspect of the economic calculation, and methods that assure the attainment and treatment of data relative to the alternative projects are necessary. Data includes information on definition and project cost estimation, on definition and estimate of benefits and advantages of the project, and on relationships and interdependences that affect or will be affected by the projects. The treatment of these data makes the measurement, evaluation, and comparison of the economic results of alternative projects possible.

With respect to the macronomics of energy, there is much interesting work in the field of modeling, mainly in as concerns analysis of the demand, in connection with the evolution of economic activity and diverse technological and social factors. The econometrics method allows one to obtain: (1) a detailed analysis of the energy demand in the level of energy uses, (2) a calculation of the final energy for each type of use, whereas the necessities of useful energy derived from, technical-economic and social indicators, (3) a construction of scenarios to take into account the evolution of all non-estimated factors or those whose evolution is bound to political, economic, or energy options, and (4) a consideration of the scenarios in terms of useful and final energy and an extension of the macroeconomic models to better represent the energy demand.

1.1.2.3 Energy Savings and Efficient Use of Energy

Energy conservation refers to all those conducive actions taken to achieve a more effective use of finite energy resources. This includes rationalization of the use of energy by means of elimination of current waste and an increase in the efficiency in the use of energy. This is achieved by reducing specific energy consumption, without sacrificing the quality of life and using all possibilities to do this, even substituting one energy form for another. The objective of energy conservation is to optimize the global relationship between energy consumption and economic growth.

It is clear that there exists the possibility of sustaining an economic growth process with a smaller consumption of energy, or in other words, energy resources can be used more efficiently by applying measures that are attainable from the economic point of view, especially with high prices of energy, and are acceptable and even convenient from the ecological point of view.

Energy conservation can be generally achieved in three stages. The first stage corresponds to the elimination of energy waste, which can be achieved with minimum investment, using existent facilities appropriately. The second level corresponds to the modification of existent facilities to improve their energy efficiency. The third stage corresponds to the development of new technologies that can enable less energy consumption *per* unit of produced product.

Energy conservation can be considered an alternative source of energy, since its implementation allows reduction of the energy consumption necessary for a certain activity, without implying a reduction of the economic activity or of the quality of life.

Two examples of technologies are of special interest from the point of view of more efficient use of energy: the combined use of electric power and heat, called cogeneration, and obtaining thermal energy at a relatively low degree by means of the use of a smaller amount of energy of a higher degree, using a heat pump. There are many technologies that can be applied to energy conservation in various sectors, including transport, industry, commercial and residential, among others.

In order to develop economic studies of different strategies of energy conservation, the costs of the saving of a certain quantity of energy obtained by means of the conservation measures are compared with the cost of the energy that would be necessary should these conservation measures not be carried out.

1.1.2.4 Energy and the Environment

Another important aspect to consider is the relationship between energy consumption and preservation of the environment. Energy use is essential to satisfy human necessities. These necessities change substantially with time and humanity evolved parallel to the moderate growth of its energy consumption until the Industrial Revolution and with an actual growing energy consumption. The speed and amplitude of this development, as well as accumulative effects lead to surpassing certain limits that this consumption pattern imposes on industrial civilization endanger human survival and that of the Earth. For the first time in history, human activity can destroy the fragile essential ecological balance necessary for life reproduction, and polluting waste perturbs the cycle of bio-geo-chemicals and the risks of occurrence of accidents with massive consequences increases continuously.

The main environmental risks are intimately associated with the increase in energy consumption and derive from carbonic anhydride emissions, nitrogen oxides and sulfur, methane, chlorofluorocarbons (CFC), acid rain, greenhouse gases, *etc.*, or the risk of accidents of spills of petroleum on land and in the sea and accidents in nuclear reactors. Moreover, the elimination of the problems of residual products and of dismantling of the reactors after their useful life and the dangers of contamination associated with the use of the nuclear energy in general are further environmental risks.

The engineering proposals will be sustained in an appropriate use of energy in order to mitigate the noxious effects of the polluting residues of energy resources, mainly those of petrochemical origin.

1.1.2.5 Energy Policy

The establishment of an energy policy where various political actions are analyzed, and where norms, financial outlines, institutions, and technologies necessary to achieve a sustainable development are given, is very important. Among the main aspects to consider in the energy policy in order to ensure sustainable development are (1) to promote the preservation and improvement of environment, (2) to incorporate in the political constitution the necessary precepts that regulate the use and conservation of energy resources (mainly the renewable ones), (3) to make transparent the costs of the different energies, (4) to implant norms in order to regulate energy markets to assure a diversity of primary sources in the medium term, (5) to establish a bank of energy information of public character, and (6) to carry out a strategic plan that considers long-term energy perspectives.

The policy of sustainable development should: allow a global vision of human activities, consider the bond between energy consumption and environmental contamination, consider cultural and geographic diversity, evaluate the carried out efforts justly, foresee what is possible to do in the future, and be sustained on solid scientific and technical bases. The energy policy should also consider the abundant resources of renewable energy resources that have a smaller environmental impact.

Concerning the climatic change problem, it will be possible to achieve a reduction in the emission of greenhouse gases by means of two main actions that will be feasible in the medium and the long term. The first one is to sequestrate very important amounts of these gases by means, for instance, of preservation and incrementation of forest areas. The second, it is to take advantage of renewable energy sources. This last option has solid technological routes that lead to the proposed goal.

It is clear that there is a necessity for a complete revision of the technical and economic potentials of renewable energy resources to be able to make more precise decisions of energy politics regarding energy and the environment, which can lead towards a sustainable development.

1.2 Overview of World Energy

1.2.1 World Primary Energy Production and Consumption

The International Energy Annual (2006) presents information and trends on world energy production and consumption of petroleum, natural gas, coal, and electricity, and carbon dioxide emissions from the consumption and flaring of fossil fuels.

Between 1996 and 2006, the world's total output of primary energy petroleum, natural gas, coal, and electric power (hydro, nuclear, geothermal, solar, wind, and biomass) increased at an average annual rate of 2.3 %.

In 2006, petroleum (crude oil and natural gas plant liquids) continued to be the world's most important primary energy source, accounting for 35.9 % of world primary energy production. During the 1996 and 2006 period, petroleum production increased by 11.7 million barrels *per* day, or 16.9 %, rising from 69.5 to 81.3 million barrels *per* day. Coal was the second primary energy source in 2006, accounting for 27.4 % of world primary energy production. World coal production totaled 6.8 billion short tons. Natural gas, the third primary energy source, accounted for 22.8 % of world primary energy production in 2006. Production of dry natural gas was 3 trillion m^3.

Hydro, nuclear, and other (geothermal, solar, wind, and wood and waste) electric power generation ranked fourth, fifth, and sixth, respectively, as primary energy sources in 2006, accounting for 6.3, 5.9, and 1.0 %, respectively, of world primary energy production. Together they accounted for a combined total of 6.1 trillion kWh.

In 2006, the US, China, and Russia were the leading producers and consumers of world energy. These three countries produced 41 % and consumed 43 % of the world's total energy. The US, China, Russia, Saudi Arabia, and Canada were the world's five largest producers of energy in 2006, supplying 50.3 % of the world's total energy. Iran, India, Australia, Mexico, and Norway together supplied an additional 12.2 % of the world's total energy.

The US, China, Russia, Japan, and India were the world's five largest consumers of primary energy in 2006, accounting for 51.8 % of world energy consumption. They were followed by Germany, Canada, France, the UK, and Brazil, which together accounted for an additional 12.6 % of world energy consumption

1.2.1.1 Petroleum

Saudi Arabia, Russia, and the US were the largest producers of petroleum in 2006. Together, they produced 33.3 % of the world's petroleum. Production from Iran and Mexico accounted for an additional 9.6 %. In 2006, the US consumed 20.7 million barrels *per* day of petroleum (24 % of world consumption). China and Japan were second and third in consumption, with 7.2 and 5.2 million barrels *per* day, respectively, followed by Russia and Germany.

1.2.1.2 Natural Gas

World production of dry natural gas increased by 0.6 trillion m^3, or at an average annual rate of 2.4 %, over the period from 1996 to 2006. Russia was the leading producer in 2006 with 0.66 trillion m^3, followed by the US with 0.53 trillion m^3. Together these two countries produced 40 % of the world's total. Canada was third in production with 0.18 trillion m^3, followed by Iran and Norway, with 0.10 and 0.09 trillion m^3, respectively. These three countries accounted for 13 % of the world total.

In 2006, the US, which was the leading consumer of dry natural gas at 0.62 trillion m^3, and Russia, second at 0.47 trillion m^3, together accounted for 37 % of world consumption. Iran ranked third in consumption, with 0.11 trillion m^3, followed by Germany and Canada, at 0.10 and 0.94 trillion m^3, respectively.

1.2.1.3 Coal

Coal production increased by 1.7 billion short tons between 1996 and 2006, or at an average annual rate of 2.9 %. China was the leading producer in 2006 at 2.6 billion short tons. The US was the second leading producer in 2006 with 1.2 billion short tons. India ranked third with 499 million short tons, followed by Australia with 420 million short tons and Russia with 323 million short tons. Together these five countries accounted for 74 % of world coal production in 2006.

China was also the largest consumer of coal in 2006, using 2.6 billion short tons, followed by the US with a consumption of 1.1 billion short tons; India, Germany, and Russia together accounted for 71 % of world coal consumption.

1.2.1.4 Hydroelectric Power

Between 1996 and 2006 hydroelectric power generation increased by 503 billion kWh at an average annual rate of 1.9 %. China, Canada, Brazil, the US, and Russia were the five largest producers of hydroelectric power in 2006. Their combined hydroelectric power generation accounted for 39 % of the world total. China led the world with 431 billion kWh, Canada was second with 352 billion kWh, Brazil was third with 345 billion kWh, and the US was fourth with 289 billion kWh, followed by Russia with 174 billion kWh.

1.2.1.5 Nuclear Electric Power

Nuclear electric power generation increased by 369 billion kWh between 1996 and 2006, or at an average annual rate of 1.5 %. The US led the world in nuclear electric power generation in 2006 with 787 billion kWh, France was second with 428 billion kWh, and Japan third with 288 billion kWh. In 2006, these three countries generated

57 % of the world's nuclear electric power. Russia, China, and India accounted for almost two-thirds of the projected net increment in world nuclear power capacity between 2003 and 2005. In the reference case, Russia contributed 18 GW of nuclear capacity between 2003 and 2005, India 17 GW, and China 45 GW. Several OECD nations with existing nuclear programs also added new capacity, including South Korea with 14 GW, Japan with 11 GW, and Canada with 6 GW. The recent construction of new plants in the United States has added 16.6 GW.

1.2.1.6 Geothermal, Solar, Wind, and Wood and Waste Electric Power

Geothermal, solar, wind, and wood and waste electric generation power increased by 237 billion kWh between 1996 and 2006, at an average annual rate of 8.8 %. The US led the world in geothermal, solar, wind and wood and waste electric power generation in 2006 with 110 billion kWh, Germany was second with 52 billion kWh, followed by Spain with 27 billion kWh, Japan with 26 billion kWh, and Brazil with 17 billion kWh. These five countries accounted for 52 % of the world geothermal, solar, wind, and wood and waste electric power generation in 2006.

1.2.2 Energy Consumption by the End-use Sector

The different kinds of energies used in residential, commercial, and industrial sectors vary widely regionally, depending on a combination of regional factors, such as the availability of energy resources, the level of economic development, and political, social, and demographic factors (IEA 2006).

1.2.2.1 The Residential Sector

Energy use in the residential sector accounts for about 15 % of worldwide delivered energy consumption and is consumed by households, excluding transportation uses. For residential buildings, the physical size of structure is one indication of the amount of energy used by its occupants. Larger homes require more energy to provide heating, air conditioning, and lighting, and they tend to include more energy-using appliances. Smaller structures require less energy because they contain less space to be heated or cooled and typically have fewer occupants. The types and amounts of energy used by households vary from country to country, depending on the natural resources, climate, available energy infrastructure, and income levels.

1.2.2.2 The Commercial Sector

The need for services such as health, education, financial and government services increases as populations increase. The commercial sector, or services sector, consists of many different types of buildings. A wide range of service activities are included, such as, schools, stores, restaurants, hotels, hospitals, museums, office buildings, banks, etc. Most commercial energy use occurs through supply services such as space heating, water heating, lighting, cooking, and cooling. Energy consumed for services not associated with buildings, such as for traffic lights and city water and sewer services, is also included as commercial sector energy use. Economic growth also determines the degree to which additional activities are offered and utilized in the commercial sector.

Slow population growth in most industrialized countries contributes to slower rates of increase in the commercial energy demand. In addition, continued efficiency improvements are projected to moderate the growth of energy demand, as energy-using equipment is replaced with newer equipment. Conversely, strong economic growth is expected to include continued growth in business activity, with its associated energy use. Among the industrialized countries, the US is the largest consumer of commercially delivered energy.

1.2.2.3 The Industrial Sector

The industrial sector include a very diverse group of industries as manufacturing, agriculture, mining and construction, and a wide range of activities, such as process and assembly uses, space conditioning, and lighting. Industrial sector energy demand varies across regions and countries, based on the level of economic activity, technological development, and population, among other factors. Industrialized economies generally have more energy-efficient industry than non-industrialized countries, whose economies generally have higher industrial energy consumption relative to the GDP. On average, the ratio is almost 40 % higher in non-industrialized countries (UN 2008).

1.2.2.4 The Transportation Sector

Energy use in the transportation sector includes the energy consumed in moving people and goods by road, rail, air, water, and pipeline. The road transport component includes light-duty vehicles, such as automobiles, sport utility vehicles, small trucks, and motorbikes, as well as heavy-duty vehicles, such as large trucks used for moving freight and buses for passenger travel. Growth in economic activity and population are the key factors that determine transportation sector energy demand. Economic growth spurs increased industrial output, which requires the movement of raw materials to manufacturing sites, as well as movement of manufactured goods to end users.

A primary factor contributing to the expected increase in energy demand for transportation is the steadily increasing demand for personal travel in both non-industrialized and industrialized economies. Increases in urbanization and personal incomes have contributed to increases in air travel and to increased motorization (more vehicles) in the growing economies. For freight transportation, trucking is expected to lead the growth in demand for transportation fuel. In addition, as trade among countries increases, the volume of freight transported by air and marine vessels is expected to increase rapidly over the projection period.

1.2.3 World Carbon Dioxide Emissions

Total world carbon dioxide (CO_2) emissions from the consumption of petroleum, natural gas, and coal, and the flaring of natural gas increased from 22.8 billion metric tons of carbon dioxide in 1996 to 29.2 billion metric tons in 2006, or by 28.0 %. The average annual growth rate of carbon dioxide emissions over the period was 2.5 % (China, the US, Russia, India, and Japan were the largest sources of carbon dioxide emissions from the consumption and flaring of fossil fuels in 2006, producing 55 % of the world total). The next five leading producers of carbon dioxide emissions from the consumption and flaring of fossil fuels were Germany, Canada, the UK, South Korea, and Iran, and together they produced an additional 10 % of the world total. In 2006, China's total carbon dioxide emissions from the consumption and flaring of fossil fuels were 6.0 billion metric tons of carbon dioxide, about 2 % more than the 5.9 billion metric tons produced by the US. Russia produced 1.7 billion metric tons, India 1.3 billion metric tons, and Japan 1.2 billion metric tons.

In 2006, the consumption of coal was the world's largest source of carbon dioxide emissions from the consumption and flaring of fossil fuels, accounting for 41.3 % of the total. World CO_2 emissions from the consumption of coal totaled 12.1 billion metric tons of carbon dioxide in 2006, up 42 % from the 1996 level of 8.5 billion metric tons. China and the US were the two largest producers of (CO_2) from the consumption of coal in 2006, accounting for 41 and 18 %, respectively, of the world total. India, Japan, and Russia together accounted for an additional 14 %.

Petroleum was the second source of carbon dioxide emissions from the consumption and flaring of fossil fuels in 2006, accounting for 38.4 % of the total. Between 1996 and 2006 emissions from the consumption of petroleum increased by 1.6 billion metric tons of carbon dioxide, or 17 %, rising from 9.6 to 11.2 billion metric tons. The US was the largest producer of CO_2 from the consumption of petroleum in 2006 and accounted for 23 % of the world total. China was the second largest producer, followed by Japan, Russia, and Germany, and together these four countries accounted for an additional 21 %.

Carbon dioxide emissions from the consumption and flaring of natural gas accounted for the remaining 20.2 % of CO_2 carbon dioxide emissions from the consumption and flaring of fossil fuels in 2006. Emissions from the consumption and flaring of natural gas increased from 4.7 billion metric tons of carbon dioxide in 1996 to 5.9 billion metric tons in 2006, or by 25 %. The US and Russia were the two largest producers of carbon dioxide from the consumption and flaring of natural gas in 2006 accounting for 20 and 15 %, respectively, of the world total. Iran, Japan, and Germany together accounted for an additional 10 %.

1.2.4 Energy Perspectives

World energy consumption is expected to expand by 50% in the next 20 years. World energy consumption will continue to increase strongly as a result of robust economic growth and expanding populations in the world's developing countries. Energy demand in industrialized countries is expected to grow slowly, at an average annual rate of 0.7%, whereas energy consumption in the emerging economies of non-industrialized countries is expected to expand by 2.5% *per* year. Given expectations that world oil prices will remain relatively high throughout the projection, liquid fuels are the world's slowest growing source of energy; the consumption of liquids increases at an average annual rate of 1.2%. Projected high prices for oil and natural gas, as well as rising concerns about the environmental impacts of fossil fuel use, improve the prospects for renewable energy sources. Worldwide, coal consumption is projected to increase by 2.0% *per* year. The cost of coal is comparatively low relative to the cost of liquids and natural gas, and abundant resources in large energy-consuming countries (including China, India, and the US) make coal an economical fuel choice. The projected coal consumption decrease in the majority of industrialized countries is due to either the slow growth rate of coal, the electricity demand growth being slow, and natural gas, nuclear power, and renewable being likely to be used for electricity generation rather than coal. Although liquid fuels and other petroleum products are expected to remain important sources of energy, natural gas remains an important fuel for electricity generation worldwide because it is more efficient and less carbon intensive than other fossil fuels. The use of hydroelectricity and other grid-connected renewable energy sources continues to expand, with consumption projected to increase by 2.1% *per* year.

Natural gas and coal, which are currently are the fastest growing fuel sources for electricity generation worldwide, continue to lead the increase in fuel use in the electric power sector. The strongest growth in electricity generation is projected for non-industrialized countries, increasing by 4.0%, as rising standards of living increase the demand for home appliances and the expansion of commercial services, including hospitals, office buildings, *etc.* In industrialized nations, where infrastructures are well established and population growth is relatively slow, a much slower growth in generation is expected, *i.e.*, 1.3%.

Because natural gas is an efficient fuel for electric power generation and produces less carbon dioxide than coal or petroleum products, it is an attractive choice in many nations.

Rising fossil fuel prices, energy security, and greenhouse gas emissions support the development of new nuclear energy generating capacities. Most expansion of installed nuclear power capacity is expected in non-industrialized countries.

There is still considerable uncertainty about the future of nuclear power, however, and a number of issues may slow the development of new nuclear power plants. Plant safety, radioactive waste disposal, and the proliferation of nuclear weapons, which continue to raise public concerns in many countries, may hinder plans for new installations, and high capital and maintenance costs may keep some countries from expanding their nuclear power programs.

Renewable fuels are the fastest growing source of energy. Higher fossil fuel prices, particularly for natural gas in the electric power sector, along with government policies and programs supporting renewable energy, allow renewable fuels to compete economically. The use of hydroelectricity and other grid-connected renewable energy sources continues to expand, with consumption projected to increase by an average of 2.1 % *per* year. Much of the growth in renewable energy consumption is projected to come from mid-scale to large-scale hydroelectric facilities in non-industrialized countries. Most of the increase in renewable energy consumption in industrialized countries is expected to come from non-hydroelectric resources, such as wind, solar, geothermal, municipal solid waste, and biomass. The European Union (EU) has set a target of increasing the renewable energy share to 20 % of gross domestic energy consumption by 2020, including a mandatory minimum of 10 % for bio fuels. Most EU member countries offer incentives for renewable energy production, including subsidies and grants for capital investments and premium prices for generation from renewable sources. Installation of wind-powered generating capacity has been particularly successful in Germany and Spain.

1.3 Air Conditioning Needs

Environmental conditions play an important role in the development of human activities. The relationship between humidity, temperature and wind velocity should create particular conditions on physiological well-being, which depends on the geographic location, specific activity, and in many cases, on cultural, social, and economic factors. Outside this area of well-being, it will be required in particular for each case of heating or cooling, the elimination or addition of humidity, as well as a control of the velocity of the air.

Man has created his own habitat to protect himself against inclement weather, where structures and appropriate building materials must be used, as well as adequate clothing. Throughout history mankind has responded to the air conditioning problem by means of vernacular architecture, using available materials that have allowed conservation or dissipation of thermal energy, reducing the requirements of conventional refrigeration.

Population growth, emigration toward urban zones, abandonment of agricultural activity, changes in design and in construction materials and architectural structures have contributed to the creation of microclimates where important effects have been had on the augmentation of temperature, humidity and in modifications of the patterns of the wind.

The abuse of the use of materials with high thermal inertia, the indiscriminate use of glass like structural material, where to avoid the introduction of solar radiation – filters have been placed that diminish brightness and increase electricity consumption for illumination – the absence of natural ventilation, among other things, have increased the temperature in the interior of rooms. Another factor has been the increase of the electric equipment in the interior, computers and their peripherals, fans, coffee, and a great diversity of appliances that dissipate heat, which is reinstated to the interior.

The external factors, such as the amount of gases and vapors of water, products of the combustion of hydrocarbons, and in the transport and industrial sectors, the greenhouse gases emission (global heating) and the decrease of ozone gas in the stratospheric atmosphere layer due to the emission of certain refrigerants have caused increases of temperature in certain regions of the world, with alarming consequences, such as an increase in pluvial precipitation, thaw, *etc.* It is also necessary to mention the climatic changes originated by desertification and the increase of the use of soil for agricultural and urban purposes. Additionally an inadequate handling of air conditioning facilities has resulted in inadequate operation and bigger energy consumption.

All of the above-mentioned factors have caused an important increase in the requirement of cooling, of refrigeration, and air conditioning, which are highly intensive in electricity with more than 15 % of what is generated in the whole world.

In order to diminish the energy requirements for air conditioning, very diverse strategies exist, among them, from the industrial point of view, efficient equipment production and the integration of appropriate, environmentally-friendly refrigerants. In the domestic and services sectors, for example, new equipment, thermal isolation of walls and roofs, decrease in the heat generated to the interior by various domestic electric appliances, and a more appropriate handling of air conditioning systems.

Another strategy consists of the diversification of the use of the conventional systems of refrigeration, based on mechanical compression, to other cooling methods based on the use of thermal energy, such as the sorption refrigeration cycles in its different processes and configurations.

Sorption refrigeration systems such as absorption and adsorption ones operate with thermal energy of a low level of temperature (90–200 °C), derived from the use of thermal solar energy conversion, the waste heat of industrial and agricultural activities, biogas combustion, among thermal sources. There exists a great potential in the recovery of dissipated heat from fuel cells, in particular the proton exchange membrane type, which promises to be an energy technology with big perspectives and where it is possible to obtain a cogeneration process where electric power and refrigeration generation is simultaneous using the energy dissipated by their own fuel cell for the thermal operation of air conditioning system.

1.4 Cogeneration Systems

Today, energy is perhaps the driving force of most economies in the world. Electric power is essential for lighting and operating equipment and appliances used in

commercial, institutional, and industrial facilities. Conventional generation contributes to two-thirds of all fuel used to make electricity, which is generally wasted
by venting unused thermal energy from power generation equipment into the air or
discharging it into water streams.

Cogeneration can be defined as the simultaneous production of electric power
and useful heat from the burning of a single fuel. This technique of combined heat
and power production has been applied successfully in industrial and tertiary sectors; the energy resources are used more efficiently, which creates opportunities
for reductions in both purchased energy costs and in environmental impact. This
occurs mainly because of efficient technology levels and the guarantee of available electricity and low level environmental impact.

Integrated systems for combined heat and power significantly increase efficiency of energy utilization, up to 80%, by using thermal energy from power
generation equipment for cooling, heating, and humidity control systems. Figure 1.1 shows that a typical cogeneration system can reduce energy requirements
by close to 20% compared to separate production of heat and power. For 170 units
of input fuel, the cogeneration system converts 130 units to useful energy of which
50 units are electricity and 80 units are for steam or hot water. Traditional separate
heat and power components require 215 units of energy to accomplish the same
end use tasks.

COGENERATION SYSTEMS

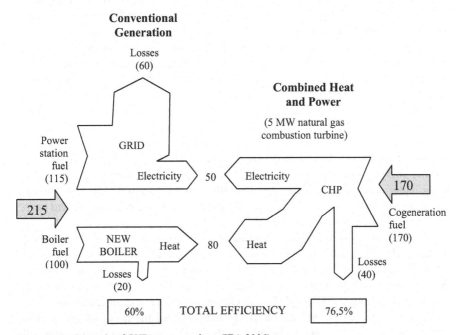

Figure 1.1 Example of CHP energy savings (IEA 2006)

A cogeneration system has the potential to dramatically reduce industrial sector carbon and air pollutant emissions and increase source energy efficiency. Industrial applications of cogeneration systems have been around for decades, producing electricity and by-product thermal energy onsite, and converting 75 % or more of the input fuel into useable energy. Typically, cogeneration systems operate by generating hot water or steam from the recovered waste heat and using it for process heating, but it also can be directed to an absorption chiller where it can provide process or space cooling. These applications are also known as cooling, heating, and power (CHP).

1.4.1 Centralized versus Distributed Power Generation

The traditional model of electric power generation and delivery is based on the construction of large, centrally-located power plants. "Central" means that a power plant is located on a hub surrounded by major electrical load centers. For instance, a power plant may be located close to a city to serve the electrical loads in the city and its suburbs or a plant may be located at the midpoint of a triangle formed by three cities.

Power must be transferred from a centrally-located plant to the users. This transfer is accomplished through an electricity grid that consists of high-voltage transmission systems and low-voltage distribution systems. High-voltage transmission systems carry electricity from the power plants to sub-stations. At the sub-stations, the high-voltage electricity is transformed into low-voltage electricity and distributed to individual customers.

Inefficiencies are associated with the traditional method of electric power generation and delivery. Figure 1.2 illustrates the losses inherent to the generation and delivery of electric power in traditional centralized power plants and in distributed power plants.

Traditional power plants convert about 30 % of the fuel's available energy into electric power, and highly efficient, distributed power plants convert over 50 % of available energy into electric power (Hardy 2003). The majority of the energy content of the fuel is lost at the power plant through the discharge of waste heat. Further energy losses occur in the transmission and distribution of electric power to the individual user. Inefficiencies and pollution issues associated with conventional power plants provide the impetus for new developments in "onsite and near-site" power generation.

The traditional structure of the electrical utility market has resulted in a relatively small number of electric utilities. However, current technology permits development of smaller, less expensive power plants, bringing in new, independent producers. Competition from these independent producers along with the re-thinking of existing regulations has affected the conventional structure of electric utilities.

The restructuring of the electric utility industry and the development of new "onsite and near-site" power generation technologies have opened up new possi-

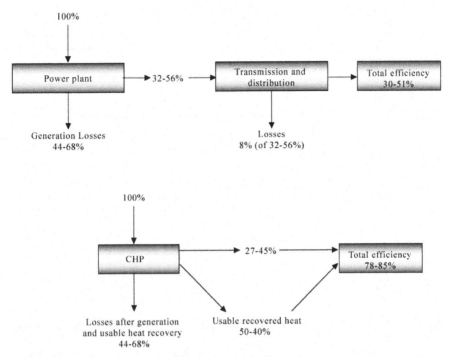

Figure 1.2 Centralized *versus* distributed power generation

bilities for commercial, public, industrial facilities, building complexes, and communities to generate and sell power. Competitive forces have created new challenges as well as opportunities for companies that can anticipate technological needs and emerging market trends.

Distributed power generation using cogeneration systems has the potential to reduce carbon and air pollutant emissions and to increase resource energy efficiency dramatically. Cogeneration systems produce both electric or shaft power and useable thermal energy onsite or near site, converting as much as 80% of the fuel into useable energy. A higher efficiency in energy conversion means that less fuel is necessary to meet energy demands. Also, onsite power generation reduces the load on the existing electricity grid, resulting in better power quality and reliability. Additionally, cogeneration systems include values such as variable fuel requirements, enhanced energy-security, and improved indoor air quality.

1.4.2 Cogeneration Technologies

Cogeneration systems utilizing internal combustion engines (Otto and Diesel versions), steam turbines, and gas turbines in open cycle are the most utilized technologies worldwide. However, some emerging technologies have become current

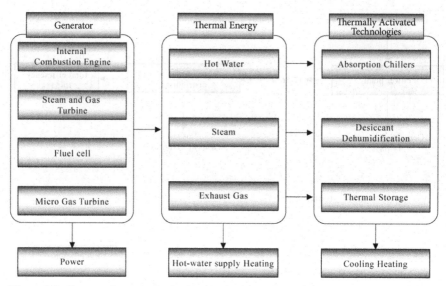

Figure 1.3 Cogeneration technologies

applications, *e.g.*, micro gas turbines in closed cycles and fuel cell systems (see Figure 1.3).

The most often cogeneration systems encountered are a gas turbine generator or reciprocating engine generator coupled with a waste heat recovery boiler, or a steam boiler coupled with a steam turbine generator. The main difference between the two types of systems is the order in which the electricity is obtained. The gas turbine and reciprocating engine first produce electricity, then the hot exhaust gases are sent to the waste heat boiler to generate steam, a process known as a topping cycle. When a boiler produces steam first and then some (or all) of that steam is sent to a steam turbine to generate electricity, the process is considered a bottoming cycle.

Depending upon the nature of the installation using these cogeneration systems, each has its advantages. The gas turbine and reciprocating engine systems are much better for new installations. The amount of power produced for a given heat demand is superior to that of the boiler/steam turbine system. For retrofit applications, where a boiler is already installed and running, the steam turbine may be ideally suited. Many installations generate steam at a higher pressure than necessary then throttle the steam to a lower pressure before it is sent to process. Replacing the pressure reducing valve with a steam turbine recovers the energy wasted in the throttling process and converts it to electricity. Moreover, since steam turbines are relatively inexpensive, the first cost is minimal.

The main advantage of operating fuel cells in a cogeneration mode is that the system consumes less fuel than would be required to produce the same thermal and electrical energy in separate processes, because of efficient technology levels. The guarantee of the available electricity and low level environmental impacts are further advantages (Rosen 1990).

Table 1.1 Waste heat characteristics of power generation technologies (Onsite Sycom 1999)

	Usable temp. for co-generation (°C)	Cogeneration output (J/kWh)	Uses for heat recovery
Diesel engine	80–500	3,400	Hot water, LP steam, district heating
Natural gas engine	150–250	1,000–5,000	Hot water, LP steam, district heating
Gas turbine	250–600	3,400–12,000	Direct heat, hot water, LP-HP steam, district heating
Micro-turbine	200–350	4,000–15,000	Direct heat, hot water, LP steam
Fuel cell	60–370	500–3,700	Hot water, LP steam

LP: low pressure, HP: high pressure

1.4.3 Heat Recovery in Cogeneration Systems

Electrical and shaft power generation efficiencies have attained maximum values of 50 % for internal combustion engines, 60 % for combustion turbines (combined cycle), 30 % for micro gas turbines, and 70 % for fuel cells (Onsite Sycom 1999). Most power generation components falling into these categories do not reach the upper level efficiencies of these technologies. Components such as micro gas turbines that convert 30 % of the input fuel into electrical or shaft power fail to harness 70 % of the available energy source. Energy that is not converted to electrical power or shaft power is typically rejected from the process in the form of waste heat. The task of converting waste heat to useful energy is called heat recovery and is primarily accomplished through the use of heat exchanger devices such as heat recovery steam generators (HRSG), water heaters, or air heaters.

The characteristics of waste heat generated in combustion turbines, internal combustion engines, and fuel cells directly affect the efficacy with which useful energy is recovered for additional processes. Some of the characteristics of the waste heat generated by these power generation technologies are presented in Table 1.1.

Waste heat is typically produced in the form of hot exhaust gases, process steam, and process liquids/solids. In combustion turbines and internal combustion engines, heat is rejected in the combustion exhaust and the coolant. Fuel cells reject heat in the form of hot water or steam.

We can classify recovered heat as low-temperature ($<230\,°C$), medium temperature (230–$650\,°C$), or high-temperature ($>650\,°C$) (Shah 1997). Recovered heat that is utilized in the power generation process is internal heat recovery, and recovered heat that is used for other processes is external heat recovery. Combustion pre-heaters, turbochargers, and recuperators are examples of internal heat recovery components. Heat recovery steam generators, absorption chillers, and desiccant systems are examples of external heat recovery components.

1.4.4   Cogeneration System Selections

Normally, cogeneration applications are geared to accomplishing two loads: an electrical load and a thermal one. Too often this load following results in difficulties in both sizing and operation of the cogeneration equipment, and limits the operational capacity of this equipment to the smallest load to be followed. Therefore, the potential savings of a cogeneration system, as related to its capacity, is restricted to this smallest load.

One the most important factors affecting the choice is the magnitude of each type of load, both thermal and electrical (see Table 1.2). If either of these is relatively low, or even non-existent, then a cogeneration system is obviously not an option. In most cases, no thermal load exists. Only in the rarest of circumstances would it be economically feasible to generate power while not recovering any thermal energy. If it does appear that pure power generation is an economic possibility, a detailed study of the power company rate structure that serves the facility should be performed. It is likely that changing to another rate structure would lower electrical costs enough to make the pure power generation option economically undesirable.

Another criterion is the size of the electrical and thermal loads relative to each other. This should not be confused with the first criteria. We are assuming that the magnitude of each type of load is sufficient to consider a cogeneration system. For high electrical usage *vs.* thermal usage, a system with a higher electrical efficiency is desirable, for example, a reciprocating engine generator. If the opposite is true, the thermal load outpaces the electrical load, and then a steam turbine would better suit the application. Finally, if both are relatively equal, then a gas turbine system might be the initial system to analyze.

The relative magnitudes of the thermal and electrical loads are not the only criteria, but also the time dependent nature of each load. Loads that vary considerably with respect to time can cause undesirable effects on certain systems, much

Table 1.2 Classifications of cogeneration systems by size range
(http://www.chpcentermw.org/presentations/WI-Focus-on-Energy-Presentation-05212003.pdf)

Systems designation	Size range	Comments
Mega	50 to 100+ MWe	Very large industrial Usually multiple smaller units Custom engineered systems
Large	10s of MWe	Industrial and large commercial Usually multiple smaller units Custom engineered systems
Mid	10s of kWe to Several MWe	Commercial and light industrial Single to multiple units Potential packaged units
Micro	<60 kWe	Small commercial and residential Appliance-like

more so than on others when a load following operational strategy is used. A reciprocating engine generator responds much better to changing loads than a gas turbine does, not only in terms of efficiency, but also reliability. Steam turbines can match loads well by simply throttling the steam flow through the turbine.

An important consideration when choosing a cogeneration system is what type of fuel is most readily available. For almost every fuel, there is a system capable of using it. Gaseous fuel, such as natural gas, is most commonly used in gas turbines, but it is also used in natural gas fired reciprocating engines. Fuels such as No. 2 and No. 6 oils are burned in reciprocating engines, and No. 2 oil is used as a backup fuel for gas turbines. Solid fuels, such as coal and biomass, are exclusively used in the Rankine cycle. Except for the solid fuels, any fuel can be used in any system, so a certain amount of flexibility exists. However, using a fuel other than the ideal will cause increased operating costs and decreased equipment life.

The type of industry choosing to cogenerate will often determine the fuel, and thereby cogeneration system to be used. The paper industry, which generates a great deal of biomass and chemical by product fuel, generally opts for a Rankine cycle system to utilize the readily available fuel source. The huge boilers burn both bark removed from the incoming logs and chemical liquor generated in the pulp making process. Similarly, the petroleum industry most often relies on fuel oil as a heat source. Because of the available supply and low cost associated with using one's own fuel oil, it makes excellent economic sense to do so. However, for those industries that do not generate a fuel source in their production process, natural gas is often the best choice, due to the low cost, high efficiency, ease of transport, and low capital cost of the storage and distribution equipment (Bretton 1997).

Pollution concerns have become particularly important in recent years, especially in heavily populated areas. Gaseous fuels tend to have the lowest emissions, followed by fuel oils, and finally solid fuels. However, in large industrial locations using solid fuels, exhaust stacks are equipped with scrubbers or precipitators to remove particulate matter or other pollutants from exhaust, thus minimizing pollution concerns. It should be borne in mind that these scrubbers add considerable cost to the overall system, and any economic analysis should include the additional capital outlay. Heavier grade fuel oils can have a high sulfur content, and unless special steps are taken, sulfur emissions can be considerable. Low sulfur oils are available, but again at a higher cost. Finally, the efficient operation of each of the systems will minimize the pollutants generated in the combustion process. If the combustion process for any of the systems is poorly managed (through combustion air, *etc.*), or maintenance is not performed at required intervals, pollutants can increase dramatically.

The physical space available for a cogeneration system will often affect which type of equipment is used. Gas turbines and reciprocating engine generators are compact, packaged units which are simply dropped into place, attached to the fuel, steam, and electrical systems, and started. Steam turbine systems usually require more on-site preparation, but only because "drop-in" packaged units do not exist. For completely new systems, steam turbine cogeneration systems are the most expensive, due to the high cost of the boilers, condensers, and other associated equipment required for operation.

The operational cost is a key factor in choosing a cogeneration system. Systems that have high fuel, maintenance, or supervisory costs will undermine any savings gained from cogenerating. Generally speaking, reciprocating engine generators have the highest operating costs, in terms of downtime and preventive maintenance, due to the high number of moving parts in the system. Steam turbine systems have lower maintenance costs than reciprocating engines, with gas turbines having the lowest costs of all.

Some general guidelines have been developed through experience with regard to selecting a prime mover (Dyer 1991). Specifically, reciprocating engines, micro gas turbines, and fuel cells tend to prosper in smaller systems (micro and mid systems), up to 3,000 kW, or systems where a peak shaving operational strategy is used (because of the relatively short operational time). Gas turbines perform best in moderately larger applications (med and large systems), from approximately 5,000 kW up to several hundred MW. Steam turbines are ideal for the largest applications (large and mega systems) or applications where solid fuel is used, because the large boilers that use this type of fuel produce enough steam to allow for huge extraction turbines to produce sizable amounts of electricity. Steam turbines will also perform well in any situation in which steam is required at different pressures.

1.5 Cogeneration Fuel Cells – Sorption Air Conditioning Systems

1.5.1 Trigeneration

The simultaneous use of energy allows one to achieve high levels of energy efficiency, lower CO_2 emissions, a security of supply, as well as lower losses. Cogeneration is among different kinds of technologies that allow the waste heat utilization for power generation, where electricity and heat are produced simultaneously. If some cooling type is required and this is produced by the same energy source, this process is known as trigeneration (electricity, heat, and cold). Figure 1.4 shows a trigeneration system schematically.

The trigeneration process increases the energy efficiency due to better utilization of waste heat into cooling power. If sorption refrigeration systems are integrated the environmental impact is reduced due to the use of natural refrigerants (ammonia, water, methylamine, ammonium nitrate, alcohol, etc.) The trigeneration plant can be evaluated as a cogeneration plant, considering all the heat used in producing cold.

This cooling can be done through sorption (absorption or adsorption refrigeration) cycles. These systems are adapted in order to recover industrial and commercial waste heat, hot liquid or hot gas, and steam, to provide cold for air condi-

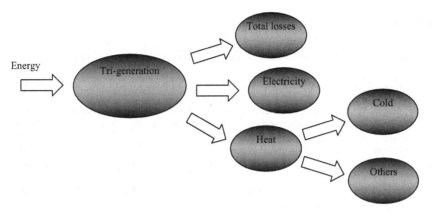

Figure 1.4 A trigeneration system

tioning or low temperature processes, as it is possible to achieve high rates of performance using residual thermal flows at relatively low temperature.

These sorption systems can be operated with thermal residual flows with a temperature range from 60–80 °C and low pressure steam, or up to 150 °C, if a double effect configuration is considered. In the case of gaseous flow, we need minimum temperatures of the order of 250 °C, due to the need for intermediate heat exchange circuit in order to generate hot water at a temperature up to 120 °C.

To generate cooling power for air conditioning application using sorption refrigeration cycles, a heat source with a temperature range between 80–200 °C (single and double effect) is required, depending on the technology selected.

1.5.2 Fuel Cells in the Trigeneration Process

The fuel cell is a technology with good performance when integrated into a trigeneration generation process. The chemical energy is transformed into electricity, heat, and water. These devices have high efficiency, low emission and noise, and a modular design. Their main practical applications are in the transport sector. Fuel cells are classified by the electrolyte used and the operating temperature. Molten carbonate (MC) and solid oxide (SO) fuel cells correspond to high temperature technology (650–1050 °C) and the proton exchange membrane (PEM) and direct methanol (DM) to low temperature technology (60–250 °C). The combined use of electricity and heat produced by the electrochemical reactions gives a high overall performance of 85 %. In order to optimize the efficiency of these devices, various projects are being carried out for the use waste heat for air conditioning systems in residential, commercial, and industrial sectors.

The waste heat released by a PEM fuel cell enables one to obtain hot water with temperatures up to 80 °C, which is suitable for the operation of sorption refrigeration cycles.

References

Bretton DJ (1997) Cogeneration in the new deregulated energy environment. Thesis, Georgia Institute of Technology

Cooling, Heating, and Power for Buildings: http://www.bchp.org/owner-basic.html. Accessed 30 Dec 2008

Dyer DF, Maples G (1991) Boiler efficiency improvement. Boiler Efficiency Institute, Auburn

Hardy JD (2003) A cooling, heating, and power for buildings (CHP-b). Thesis Mississippi State University

International Energy Agency (2006) World energy outlook. www.iea.org

International Energy Annual (2006) International Energy Agency, long-term historical international energy statistics

Onsite Sycom (1999) Review of CHP technologies. US Department of Energy, Office of Energy Efficiency and Renewable Energy

Rosen MA (1990) Comparison based on energy and exergy analyses of the potential cogeneration efficiencies for fuel cells and other electricity generation devices. Int J Hydrogen Energy 15(4):267–274

Shah RK (1997) Recuperators, regenerators and compact heat exchangers. CRC Handbook of Energy Efficiency. CRC Press, New York http://www.chpcentermw.org/presentations/WI-Focus-on-Energy-Presentation-05212003.pdf. Accessed 30 Dec 2008

United Nations (2008) World economic situation and prospects

Chapter 2
Thermodynamics of Fuel Cells

2.1 Introduction

In this chapter the basic thermodynamic and electrochemical principles behind fuel cell operation and technology are described. The basic electrochemistry principles determining the operation of the fuel cell, the kinetics of redox reactions during the fuel cell operation, the mass and energy transport in a fuel cell, *etc.*, are described briefly to give an understanding of practical fuel cell systems. The ideal and practical operation of fuel cells and their efficiency are also described. This will provide the framework to understand the electrochemical and thermodynamic basics of the operation of fuel cells and how fuel cell performance can be influenced by the operating conditions. The influence of thermodynamic variables like pressure, temperature, and gas concentration, *etc.*, on fuel cell performance has to be analyzed and understood to predict how fuel cells interact with the systems where it is applied. Understanding the impact of these variables allows system analysis studies of a specific fuel cell application.

2.2 Thermodynamic and Electrochemical Principles

2.2.1 Electrochemical Aspects

All power generation systems require an energy balance to demonstrate the functioning of the system in detail. In a similar fashion the fuel cell system requires an energy or heat balance analysis (EG&G Services Parsons 2000). The energy balance analysis in the fuel cell should be based on energy conversion processes like power generation, electrochemical reactions, heat loss, *etc.* The energy balance analysis varies for the different types of fuel cells because the various types of electrochemical reactions occur according to the fuel cell type. The enthalpy of the reactants entering the system should match the sum of the enthalpies of the products leaving the cell, the net heat generated within the system, the dc power output from the cell, and the heat loss from the cell to its surroundings. The energy balance analysis is done by determining the fuel cell temperature at the exit by having information of the reactant composition, the temperatures, H_2 and O_2 utilization, the power produced, and the heat loss (Srinivasan 2006).

The fuel cell reaction (inverse of the electrolysis reaction) is a chemical process that can be divided into two electrochemical half-cell reactions. The most simple and common reaction encountered in fuel cells is (Atkins 1986)

$$H_2 + \tfrac{1}{2} O_2 \rightarrow H_2O \tag{2.1}$$

Analyzing from a thermodynamic point of view, the maximum work output obtained from the above reaction is related to the free-energy change of the reaction. Treating this analysis in terms of the Gibbs free energy is more useful than that in terms of the change in Helmholtz free energy, because it is more practical to carry out chemical reactions at a constant temperature and pressure rather than at constant temperature and volume. The above reaction is spontaneous and thermodynamically favored because the free energy of the products is less than that of the reactants. The standard free energy change of the fuel cell reaction is indicated by the equation

$$\Delta G = -nFE \tag{2.2}$$

Where ΔG is the free energy change, n is the number of moles of electrons involved, E is the reversible potential, and F is Faraday's constant. If the reactants and the products are in their standard states, the equation can be represented as

$$\Delta G^0 = -nFE^0 \tag{2.3}$$

The value of ΔG corresponding to (2.1) is $-229\,\text{kJ/mol}$, $n=2$, $F = 96500\,\text{C/g.mole}$ electron, and hence the calculated value of E is 1.229 V.

The enthalpy change ΔH for a fuel cell reaction indicates the entire heat released by the reaction at constant pressure. The fuel cell potential in accordance with ΔH is defined as the thermo-neutral potential, E_t,

$$\Delta H = -nFE_t \tag{2.4}$$

where E_t has a value of 1.48 V for the reaction represented by Equation 2.1.

The electrochemical reactions taking place in a fuel cell determine the ideal performance of a fuel cell; these are shown in Table 2.1 for different kinds of fuels depending on the electrochemical reactions that occur with different fuels, where CO is carbon monoxide, e^- is an electron, H_2O is water, CO_2 is carbon dioxide, H^+ is a hydrogen ion, O_2 is oxygen, CO_3^{2-} is a carbonate ion, H_2 is hydrogen, and OH^- is a hydroxyl ion.

It is very clear that from one kind of cell to another the reactions vary, and thus so do the types of fuel. The minimum temperature for optimum operating conditions varies from cell to cell. This detail will be discussed in subsequent chapters. Low to medium-temperature fuel cells such as polymer electrolyte fuel cells

(PEMFC), alkaline fuel cells (AFC), and phosphoric acid fuel cells (PAFC) are limited by the requirement of noble metal electrocatalysts for optimum reaction rates at the anode and cathode, and H_2 is the most recommended fuel. For high-temperature fuel cells such as molten carbonate fuel cells (MCFC) and solid oxide fuel cells (SOFC) the catalyst restrictions are less stringent, and the fuel types can vary. Carbon monoxide can poison a noble metal electrocatalyst such as platinum (Pt) in low-temperature fuel cells, but it serves as a potential fuel in high-temperature fuel cells where non-noble metal catalysts such as nickel (Ni), or oxides can be employed as catalysts.

The ideal performance of a fuel cell can be represented in different ways. The most commonly used practice is to define it by the Nernst potential represented as the cell voltage. The fuel cell reactions corresponding to the anode and cathode reactions and the corresponding Nernst equations (Simons *et al.* 1982, Cairns and Liebhafsky 1969) are given in Table 2.2.

The Nernst equation is a representation of the relationship between the ideal standard potential E^0 for the fuel cell reaction and the ideal equilibrium potential E at other temperatures and pressures of reactants and products. Once the ideal potential at standard conditions is known, the ideal voltage can be determined at other temperatures and pressures through the use of these equations. According to the Nernst equation for hydrogen oxidation, the ideal cell potential at a given temperature can be increased by operating the cell at higher reactant pressures. Improvements in fuel cell performance have been observed at higher pressures and temperatures. The symbol E represents the equilibrium potential, E^0 the standard potential, P the gas pressure, R the universal gas constant, F Faraday's constant and T the absolute temperature.

Table 2.1 Summary of the electrochemical reactions taking place in different fuel cells

Fuel cell type	Anode reaction	Cathode reaction
Acid fuel cell (including PEM)	$H_2 \rightarrow 2H^+ + 2e^+$	$\frac{1}{2}O_2 + 2H^+ + 2e^+ \rightarrow H_2O$
Alkaline fuel cell	$H_2 + 2(OH)^- \rightarrow 2H_2O + 2e^-$	$\frac{1}{2}O_2 + H_2O + 2e^- \rightarrow 2(OH)^-$
Oxide fuel cell	$H_2 + O^{2-} \rightarrow H_2O + 2e^-$ $CO + O^{2-} \rightarrow CO_2 + 2e^-$ $CH_4 + 4O^{2-} \rightarrow 2H_2O + CO_2 + 8e^-$	$\frac{1}{2}O_2 + 2e^- \rightarrow O^{2-}$
Molten carbonate fuel cell	$H_2 + CO_3^{2-} \rightarrow H_2O + CO_2 + 2e^-$ $CO + CO_3^{2-} \rightarrow 2CO_2 + 2e^-$	$\frac{1}{2}O_2 + CO_2 + 2e^- \rightarrow CO_3^{2-}$

Table 2.2 The relationship between fuel cell reaction and the Nernst equation

Fuel cell reaction	Nernst equation
$H_2 + \frac{1}{2}O_2 \rightarrow H_2O$	$E = E^0 + (RT/2F) \ln [P_{H2}/P_{H2O}] + (RT/2F) \ln [P_{O2}^{1/2}]$
$H_2 + \frac{1}{2}O_2 + CO_{2\,(cathode)} \rightarrow$ $H_2O + CO_{2\,(anode)}$	$E = E^0 + (RT/2F) \ln [P_{H2}/P_{H2O} (P_{CO2anode})] +$ $(RT/2F) \ln [P_{O2}^{1/2} (P_{CO2cathode})]$
$CO + \frac{1}{2}O_2 \rightarrow CO_2$	$E = E^0 + (RT/2F) \ln [P_{CO}/P_{CO2}] + (RT/2F) \ln [P_{O2}^{1/2}]$
$CH_4 + 2O_2 \rightarrow 2H_2O + CO_2$	$E = E^0 + (RT/8F) \ln [P_{CH4}/P_{H2O}^2 P_{CO2}] + (RT/8F) \ln [P_{O2}^2]$

In general in a fuel cell the reaction of H_2 and O_2 produces H_2O. When hydrocarbon fuels are involved in the anode reaction, CO_2 is also produced. For molten carbonate fuel cells CO_2 is consumed in the cathode reaction to maintain the invariant carbonate concentration in the electrolyte. Since CO_2 is generated at the anode and consumed at the cathode in MCFCs, and because the concentrations of the anode and cathode flows are not necessarily equal, the Nernst equation in Table 2.2 includes the partial pressures of CO_2 for both electrode reactions.

The ideal standard potential of an H_2/O_2 fuel cell (E^0) is 1.229 V with liquid water as the product and 1.18 V for water with gaseous product. This value is normally referred to as the oxidation potential of H_2. The potential can also be expressed as a change in Gibbs free energy for the reaction of hydrogen and oxygen. The change in Gibbs free energy increases as cell temperature decreases and the ideal potential of a cell is proportional to the change in the standard Gibbs free energy. This will be discussed in more detail in the thermodynamics sections of the other chapters.

The variation of the standard potential in a fuel cell with temperature is shown in Figure 2.1. It is very clear that the influence of temperature on the standard potential is more pronounced for high-temperature fuel cells. This case corresponds to low, medium, and high-temperature fuel cells. Hence the ideal potential is less than 1.229 V when considering the gaseous water product in a fuel cell.

The ideal and actual performance of a fuel cell is quite different, especially when one analyzes the potential current response of a fuel cell. Figure 2.2 displays the ideal and actual responses of a fuel cell. Electrical energy is obtained from a fuel cell when a current is drawn, but the actual cell potential is lowered from its equilibrium potential because of irreversible losses due to various reasons. Several factors contribute to the irreversible losses in a practical fuel cell. The losses, which are generally called polarization or over potential, originate primarily from activation polarization, ohmic polarization, and gas concentration polarization (Chase *et al.* 1985). These losses result in a cell potential for a fuel cell that is less than its ideal potential.

The first of these three major polarizations is the activation loss, which is pronounced in the low current region. In this region electronic barriers must be overcome before the advent of current and ionic flow. The activation loss is directly proportional to the increase in current flow. The activation polarization can be represented as

$$\eta_{act} = \frac{RT}{\alpha nF} \ln\left(\frac{i}{i_0}\right)$$

$$(2.5)$$

Where η_{act} is the activation polarization, R the universal gas constant, T the temperature, α the charge transfer coefficient, n the number of electrons involved, F the Faraday constant, i the current density, and i_0 the exchange current density. Activation polarization is due to the slow electrochemical reactions at the electrode surface, where the species are oxidized or reduced in a fuel cell reaction. Activation polarization is directly related to the rate at which the fuel or the oxi-

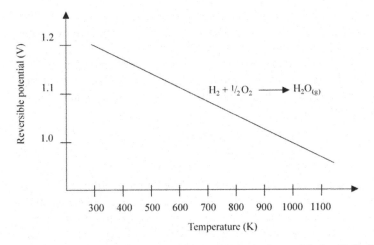

Figure 2.1 The influence of temperature on the standard potential of an H_2/O_2 fuel cell

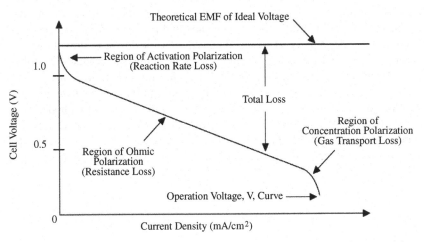

Figure 2.2 Ideal and actual performance of a fuel cell with respect to the potential current response

dant is oxidized or reduced. In the case of fuel cell reactions the activation barrier must be overcome by the reacting species.

The ohmic polarization varies proportionally to the increase in current and increases over the entire range of currents due to the constant nature of fuel cell resistance. The ohmic polarization can be represented as

$$\eta_{ohm} = iR_c \tag{2.6}$$

Where η_{ohm} is the ohmic polarization and R_c is the cell resistance.

The origin of ohmic polarization comes from the resistance to the flow of ions in the electrolyte and flow of electrons through the electrodes and the external electrical circuit. The dominant ohmic loss is in the electrolyte, which is reduced by decreasing the electrode separation, enhancing the ionic conductivity of the electrolyte and by modification of the electrolyte properties.

The concentration losses occur over the entire range of current density, but these losses become prominent at high limiting currents where it becomes difficult for gas reactant flow to reach the fuel cell reaction sites. The concentration polarization can be represented as

$$\eta_{con} = \left(\frac{RT}{nF}\right)\ln\left(1-\frac{i}{i_\Delta}\right) \tag{2.7}$$

Where η_{con} is the concentration polarization, i_L is the limiting current density. As the reactant gas is consumed at the electrode through the electrochemical reaction, there will be a potential drop due to the drop in the initial concentration of the bulk of the fluid in the surroundings. This leads to the formation of a concentration gradient in the system. Several processes are responsible for the formation of the concentration polarization. These are (1) slow diffusion of the gas phase in the electrode pores, (2) solution of reactants into the electrolyte, (3) dissolution of products out of the system, and (4) diffusion of reactants and products, from the reaction sites, through the electrolyte. At practical current densities there is slow transport of reactants to the electrochemical reaction and slow removal of products from the reaction site, which is a major contributor to the concentration polarization.

Figure 2.3 depicts the schematic representation of the various contributions to polarization losses in a fuel cell, especially those from anode and cathode. The net result of concentration polarization in current flow in a fuel cell is to increase the anode potential and to decrease the cathode potential. This will result in the reduction of the cell voltage. These polarization curves are typical for each type of fuel cell.

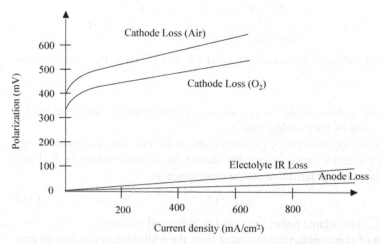

Figure 2.3 Contributions to various polarization losses coming from the anode and cathode of a fuel cell

2.2.2   Thermodynamic Principles

The effect of temperature and pressure on the cell potential may be analyzed on the basis of the Gibbs free energy variation with respect to temperature and pressure in a fuel cell. This may be written as

$$\left(\frac{\partial E}{\partial T}\right)_P = \frac{\Delta S}{nF} \tag{2.8}$$

$$\left(\frac{\partial E}{\partial P}\right)_T = -\frac{\Delta V}{nF} \tag{2.9}$$

where $-\Delta V$ is the change in volume, ΔS is the entropy change, E is the cell potential, T the temperature, P the reactant gas pressure, n the number of electrons transferred, and F Faraday's constant.

Since the entropy change for the H_2/O_2 fuel cell reaction is negative the reversible potential of the H_2/O_2 fuel cell decreases with an increase in temperature, assuming that the reaction product is liquid water. For the above reaction the volume change is negative, hence the reversible potential increases with an increase in pressure. The influence of temperature on the fuel cell voltage is shown schematically in Figure 2.4, where the fuel cell performance data from typical operating cells and the dependence of the reversible potential of H_2/O_2 fuel cells on temperature are given. The cell voltages of PEFC, PAFC, and MCFC show a strong dependence on temperature (Appleby and Foulkes 1989, Angrist 2000). The reversible potential decreases with increasing temperature, but the operating voltages of these fuel cells actually increase with an increase in operating temperature. PEFC exhibits a maximum in operating voltage. The lower operating temperature of SOFC is limited to about 1000 °C since the ohmic resistance of the solid elec-

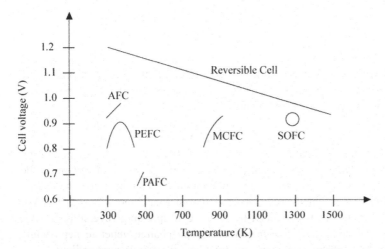

Figure 2.4   The influence of temperature on the operating voltage of different fuel cells

trolyte increases rapidly as the temperature decreases. However, advances in materials science in developing new solid oxide electrolytes and thin film solid electrolytes have succeeded in lowering the minimum operating temperature of SOFC below 1000 °C. Normally fuel cells operate at voltages considerably lower than the reversible cell voltage. The better performance is related to changes in the types of polarizations affecting the cell as the temperature varies. An increase in the operating temperature is beneficial to fuel cell performance because of the increase in reaction rate, higher mass transfer rate, and usually lower cell resistance arising from the higher ionic conductivity of the electrolyte. In addition, the CO tolerance of electrocatalysts in low-temperature fuel cells improves as the operating temperature increases.

An increase in operating pressure has several positive effects on fuel cell performance. The partial pressures of reactant gases, solubility, and mass transfer rates are higher at higher pressures. The electrolyte loss by evaporation is reduced at higher operating pressures. The system efficiency is increased by the increase in pressure. The benefits of increased pressure may be compared with the problems associated with fuel cell materials and other associated system instrumentation. Especially higher pressures increase material problems in MCFC and SOFC. Pressure differences must be minimized to prevent reactant gas leakage through the electrolyte and seals. High pressure favors carbon deposition and methane formation in the fuel gas.

The maximum electrical work obtainable in a fuel cell operating at constant temperature and pressure is given by the change in the Gibbs free energy of the electrochemical reaction,

$$W = \Delta G = -n\mathrm{F}E \qquad (2.10)$$

Where n is the number of electrons participating in the reaction, F is Faraday's constant (96,487 coulombs/g-mole electron), and E is the ideal potential of the cell. If we consider the case of reactants and products being in the standard state, then

$$\Delta G^0 = -n\mathrm{F}E^0 \qquad (2.11)$$

The overall reactions given in Table 2.2 can be used to produce both electrical energy and heat. The maximum work available from a fuel source is related to the free energy of reaction in the case of a fuel cell, whereas the enthalpy of reaction is the pertinent quantity for a heat engine, *i.e.*,

$$\Delta G = \Delta H - T\Delta S \qquad (2.12)$$

where the difference between ΔG and ΔH is proportional to the change in entropy ΔS. This entropy change is manifested in changes in the degrees of freedom for the chemical system being considered. The maximum amount of electrical energy available is ΔG as mentioned above, and the total thermal energy available is ΔH. The amount of heat that is produced by a fuel cell operating reversibly is $T\Delta S$. Reactions in fuel cells that have negative entropy change generate heat, while those with positive entropy change may extract heat from their surroundings.

Differentiating Equation 2.12 with respect to temperature or pressure, and substituting it into Equation 2.10 gives

$$\left(\frac{\partial E}{\partial T}\right)_P = \frac{\Delta S}{n\text{F}} \tag{2.13}$$

or

$$\left(\frac{\partial E}{\partial P}\right)_T = -\frac{\partial V}{n\text{F}} \tag{2.14}$$

This was demonstrated earlier in this section.

2.3 Fuel Cell Efficiency

The thermal efficiency of an energy conversion device is defined as the amount of useful energy produced relative to the change in stored chemical energy (commonly referred to as thermal energy) that is released when a fuel is reacted with an oxidant. Hence the efficiency may be defined as

$$\eta_e = \frac{\text{useful output energy}}{\Delta H} \tag{2.15}$$

Hydrogen (fuel) and oxygen (oxidant) can exist in each other's presence at room temperature, but if heated to above 500 °C and at high pressure they explode violently. The combustion reaction for these gases can be forced to occur below 500 °C in the presence of a flame, such as in a heat engine. In the case of a fuel cell, a catalyst can increase the rate of reaction of H_2 and O_2 at temperatures lower than 500 °C in the ambient of an electrolyte. In high temperature fuel cells a non-combustible reaction can occur at temperatures over 500°C because of controlled separation of the fuel and oxidant. The process taking place in a heat engine is thermal, where as the fuel cell process is electrochemical. The difference in these two processes in energy conversion is the fact behind efficiency comparison for these two systems. In the ideal case of an electrochemical energy conversion reaction such as a fuel cell the change in Gibbs free energy of the reaction is available as useful electric energy at the output of the device. The ideal efficiency of a fuel cell operating irreversibly may be stated as

$$\eta_e = \frac{\Delta G}{\Delta H} \tag{2.16}$$

The most commonly used way of expressing efficiency of a fuel cell is based on the change in the standard free energy for the cell reaction

$$H_2 + 1/2 \, O_2 \rightarrow H_2O$$

$$\Delta G^0 = G^0_{H_2O} - G^0_{H_2} - \frac{1}{2}G^0_{O_2} \tag{2.17}$$

where the product water is in liquid form. At standard conditions of reaction the chemical energy in the hydrogen/oxygen reaction is 285.8 kJ/mole and the free energy available for useful work is 237.1 kJ/mole. Thus, the thermal efficiency of an ideal fuel cell operating reversibly on pure hydrogen and oxygen at standard conditions would be

$$\eta_e = \frac{237.1}{285.8} = 0.83 \qquad (2.18)$$

The efficiency of an actual fuel cell can be expressed in terms of the ratio of the operating cell voltage to the ideal cell voltage. The actual cell voltage is less than the ideal cell voltage because of the losses associated with cell polarization and the iR loss, as discussed in the earlier section. The thermal efficiency of the fuel cell can then be written in terms of the actual cell voltage,

$$\eta_e = \frac{\text{useful output energy}}{\Delta H} = \frac{\text{useful output power}}{\left(\dfrac{\Delta G}{0.83}\right)} = \frac{V_{cell} \cdot I}{\left(\dfrac{V_{deal} \cdot I}{0.83}\right)} = \frac{V_{cell} \cdot 0.83}{V_{ideal}} \qquad (2.19)$$

As mentioned earlier, the ideal voltage of a fuel cell operating reversibly with pure hydrogen and oxygen in standard conditions is 1.229 V. Thus, the thermal efficiency of an actual fuel cell operating at a voltage of V_{cell}, based on the higher heating value of hydrogen is given by

$$\eta_e = \frac{0.83 \cdot V_{cell}}{V_{ideal}} = \frac{0.83 \cdot V_{cell}}{1.229} = 0.675 \cdot V_{cell} \qquad (2.20)$$

A fuel cell can be operated at different current densities; the corresponding cell voltage then determines the fuel cell efficiency. Decreasing the current density increases the cell voltage, thereby increasing the fuel cell efficiency. In fact, as the current density is decreased, the active cell area must be increased to obtain the desired amount of power.

2.4 Fuel Cell Operation

Fuel cell operation is influenced by various thermodynamic and electrochemical variables, such as temperature, pressure, gas concentration, reactant utilization, current density, *etc.*, which directly influence the cell potential and voltage losses. Changing the fuel cell operating parameters can have either a beneficial or a detrimental impact on fuel cell performance and on the performance of other system components. Changes in operating conditions may lower the cost of the cell, but increase the cost of the peripheral components. Generally, a compromise in the operating parameters is made to meet the required application. It is possible to obtain low system cost and achieve acceptable cell life by operating at optimum operating conditions. Operating conditions are optimized by defining specific system requirements such as power requirement level, voltage, current requirement *etc*. From this and through life cycle studies, the power, voltage, and current requirements of the fuel cell stack and individual cells are determined. It is a question of choosing an optimum cell operating point as shown by Figure 2.5 until the system requirements are satisfied. This figure shows the relation between voltage and current density and between output power and current density. For example, a design point at high current density will allow a smaller cell size at lower capital cost to be used for the stack, but a lower system efficiency results. This type of operating point would be required by a vehicle application where light weight, small volume, and efficiency are important parameters for cost effectiveness. Fuel cells capable of higher current density operation would be of special interest. Operation at a lower current density, but higher voltage would be more suitable for stationary power plant operation. Operation at a higher pressure will increase cell performance and lower cost.

Figure 2.5 displays information similar to that presented in Figure 2.4, but the former highlights another way of determining the cell design point. It is normal and seems logical to design the cell to operate at the maximum power density that peaks at a higher current density. However, operation at the higher power densities will mean operation at lower cell voltages or lower cell efficiency. Setting the operating point at the peak power density may cause instability in power control because the system will have a tendency to oscillate between higher and lower current densities around the peak. It is normal practice to operate the cell at a point towards the left side of the power density peak and at a point that yields a compromise between low operating cost and low capital cost.

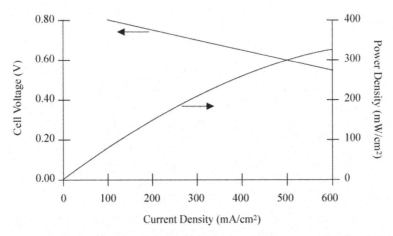

Figure 2.5 The voltage–current density and power density–current density relations for a fuel cell showing the operating conditions for a fuel cell

References

Angrist SW (2000) Direct energy conversion, 3rd edn. Allyn and Bacon, Boston

Atkins PW (1986) Physical chemistry, 3rd edn. W.H. Freeman and Company, New York

Cairns EJ, Liebhafsky HA (1969) Irreversibility caused by composition changes in fuel cells. Energy Conversion 9:63

Chase MW *et al* (1985) JANAF thermochemical tables, 3rd edn. American Chemical Society and the American Institute of Physics for the National Bureau of Standards (now National Institute of Standards and Technology)

EG&G Services Parsons Inc. (2000) Science Applications International Corporation, fuel cell hand book, 5th edn. US Department of Energy

Simons SN, King RB, Prokopius PR (1982) In: Camara EH Symposium Proceedings Fuel Cells Technology Status and Applications. Institute of Gas Technology, Chicago, p 46

Srinivasan S (2006) Fuel cells: from fundamentals to applications. Springer, New York

Chapter 3
Selected Fuel Cells for Cogeneration CHP Processes

3.1 Introduction

Before going into the details of cogeneration combined heat and power (CHP) processes, it is necessary to establish the operation principles and performance characteristics of the most common fuel cells for consideration in CHP processes. According to the phenomena governing the performance of fuel cells, it is worth noting that fuel cells are electrochemical energy conversion devices, where redox reactions occur spontaneously and the fuel and oxidant are consumed, and the electrochemical energy is transformed into electricity to produce work.

The thermodynamic reversible potentials and over-potential losses are the principal factors that control the net efficiency to convert chemical energy to electrical energy.

The fuel cell operating conditions depend on the electrochemical nature of the electricity production. Normally, activation over-potentials predominate at low current densities and they are also controlled by mass transport over-potentials at high current densities. Thus, the application of fuel cells in CHP processes implies a more detailed knowledge of the operation of every kind of fuel cell. Therefore, in this chapter, the fundamentals of the four typical fuel cells considered for CHP applications are explained, taking electrochemical operation principles and the consequent heat and electricity production into consideration. Generally, fuel cell classification is according to the type of electrolyte used and the operating temperatures.

3.2 Fuel Cell Classification

There are many types of physicochemical characteristics by which the fuel cells are classified. They can be classified by operating temperatures as low (10–80 °C), intermediate (120–200 °C), and high-temperature (650–1000 °C) fuel cells. Low-

37

temperature fuel cells are H_2/O_2-air PEMFC and CH_3OH/O_2-air DMFC. Intermediate-temperature fuel cells are H_2/O_2-air AFC and H_2/O_2-air PAFC. High temperature-fuel cells are H_2+CO/O_2-air, CH_4/O_2-air MCFC, H_2+CO/O_2-air, CH_4/O_2-air SOFC, and hybrid system MCFC-SOFC-Gas turbines (EG&G Services Parson 2000). However, the name definition of every fuel cell is due to the electrolyte used where half cell reactions occur to produce electricity. In this case, the classification is proton exchange membrane fuel cell, direct methanol fuel cell, alkaline fuel cell, phosphoric acid fuel cell, molten carbonate fuel cell, and solid oxide fuel cell.

3.2.1   The Proton Exchange Membrane Fuel Cell

The proton exchange membrane fuel cell was developed by General Electric in the 1960s for space vehicle applications. Two components are essential for the performance of this fuel cell: the electrolyte and the membrane electrodes assembly (MEA), where H^+ species are diffused from the anode to the cathode where electrochemical redox reactions take place. Protons are ions that can move through the polymer as H^+, and it is possible to consider the basic operation of this fuel cell as the functioning of a simple acid electrolyte fuel cell. A schematic representation of a single PEM fuel cell is shown in Figure 3.1.

The electrochemical reactions occurring in a PEMFC are:
At the anode

$$H_2 \rightarrow 2H^+ + 2e^- \tag{3.1}$$

At the cathode

$$\tfrac{1}{2}O_2 + 2H^+ + 2e^- \rightarrow H_2O \tag{3.2}$$

The overall reaction is

$$\tfrac{1}{2}O_2 + H_2 \rightarrow H_2O \tag{3.3}$$

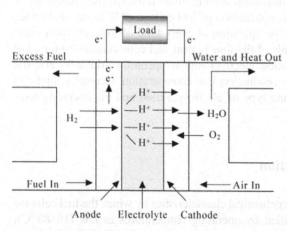

Figure 3.1 Schematic representation of the performance of PEMFC

One important performance parameter of the currently most used polymer electrolyte in PEM fuel cells (Nafion, a registered trade mark of Dupont Co.) is its capability to perform at room temperature, promoting a quick start-up of the fuel cell in a low temperature range, from 10–80 °C, which is useful for mobile, electronic, and portable applications. The integration of fuel cells in highly integrated microelectronics is possible due to the thickness of MEAs, so an appropriate and compact fuel cell can be designed for supplying energy to these kinds of devices and the heat and water formation of the fuel cell can be safely vented.

Nafion has been the most used and considered as part of the performance of PEMFC. However, this electrolyte has some inherent problems to the nature of its synthesis, for example, the water management of the polymer and the high dependence of ionic conductivity as a function of membrane hydration content. It is possible to think that the other types of fuel cells operating at low temperature are going to be used in specific applications, until the solid electrolyte can solve the above mentioned problems in addition to the specific cost of the development of catalyst technology.

The technology to develop new electrocatalysts is not just based on using less platinum as the principal catalyst used in PEMFC; it also implies the technology to develop more powerful fuel cells, improving all components of the devices. Also, the application of nanotechnology for developing new catalysts has favored the use of less catalyst by a factor of 100 from the first PEM fuel cells (using 28 mg/cm^2) to the current PEM fuel cells (using 0.2 mg/cm^2 or less). The most important advances in PEM fuel cell technology were carried out in the last 10 or 20 years. One part of this advance has been the discovery of a new technology, for example, the understanding and use of applied nanotechnology. Another advance is due to the effort of Ballard Power Systems and its partnerships with research centers (Koppel 1999).

These improvements were conducted to obtain high power fuel cells (higher power density) and a significant reduction in cost *per* kilowatt of electric power. PEMFCs have a potential and real use in automobiles, aerospace applications, portable devices (phones and laptops), and a more significant use in CHP systems (Zhang *et al.* 2006). The versatility of PEMFCs is the possibility to obtain fuel cells from few watts to hundreds of kilowatts for powering electronic and domestic devices up to vehicular and industrial systems.

3.2.1.1 Solid Electrolytes in PEMFC

There are many trademarked solid electrolytes proposed for PEMFC applications (Rozière and Jones 2001), however the most important solid electrolyte has been a thin polymer membrane called Nafion (Dupont®). Nafion has been a reference or a standard based on sulfonated fluoropolymers, in particular flouroethylene widely used in fuel cells. Nafion is a polymer similar to polytetrafluoroethylene, also called Teflon®. This polymer has played an important role in the development of fuel cells. Teflon is formed by very resistant and stable chemical bonds between fluorine and carbon atoms, showing hydrophobic nature, and this property is favorable for electrodes in fuel cells where it is necessary to drive out the water produced due to redox reactions occurring in the fuel cell operation, avoiding electrode flooding that can diminish the output power of the system. This principle is also used in other types of low and intermediate temperature fuel cells (Larminie and Dicks 2003).

An electrolyte for solid polymer fuel cell application is based on PTFE modified with sulfonated chains forming HSO_3 structures. Sulfonation of molecules is a powerful technique used in chemical processes for obtaining extraordinary materials with excellent characteristics, for example, the detergent industry. In the case of the electrolyte for fuel cells, the HSO_3 group is normally ionically bonded in the form of SO_3^- like an ionomer. SO_3^- and H^+ interact showing a strong attraction between different charged ions, promoting an agglomeration in the side chain molecules and transporting ions from one side to another through the polymer membrane. An interesting property of sulfonic acid is its hydrophyllic characteristic for absorbing water molecules.

Nafion membranes show both characteristics in one, and the effect is a solid electrolyte where hydrophilic islands exist in a hydrophobic bulk; it is expected to be responsible for the interesting properties of the Nafion membrane as electrolyte in fuel cells. However, the electrical conductivity of Nafion depends strongly on the hydrating value. For the highest conductivity conditions it is necessary to have a complete hydrated membrane showing a conductivity value of about 0.1 S/cm. When the membrane hydration decreases, the conductivity decreases much faster, and the transport of ions through the membrane is affected, causing a power loss in the performance of the fuel cell.

3.2.1.2 Electrodes and Assemblies

The performance of electrodes used in fuel cells is associated with the use of platinum. Platinum was used in the first stages of fuel cell development with a loading as high as 28 mg/cm^2. The high cost of the catalyst was balanced with the advances in platinum reduction in the electrodes to around 0.2 mg/cm^2.

The use of very small catalyst particles, (nanoparticles) allows the reduction of the amount of platinum catalyst layer in fuel cell electrodes. Probably the success of the active catalyst layer on the electrode is due to the effectiveness of the composition and preparation of the electrodes. In general, it is possible to say that the anode and cathode electrodes in a PEMFC are identical, and if the fuel and oxidant are pure, then the catalyst used can be considered as elemental platinum.

The platinum catalyst in an electrode is normally supported and spread out on carbon powder; the most widely used is XC-72 (Cabot®). The purpose of this procedure is to maintain the catalyst as an active area where contact with the reactant gases is possible and easy, and also the possibility to take out the electrogenerated electrons through the electronic conductor (carbon powder).

Two very similar methods have been reported for the preparation of electrodes. One is based on the formation of separate electrodes using carbon cloth or carbon paper with some additives like PTFE, adding catalytic ink where platinum is the active material. In this way a gas diffusion layer is formed. At the same time, the membrane is usually activated by immersion in boiling diluted hydrogen peroxide in water for 1 h or more, combining membrane immersions in sulfuric acid. The activated membrane (sulfuric acid free) and the electrodes are put together and the assembly is hot pressed at a controlled temperature around 140 °C at high pressure for less than 5 min. The product obtained is the membrane electrode assembly (MEA), which has shown excellent performance in PEMFCs.

The other method that has started being used is electrode-membrane self-assembly. This method involves the deposition of the electrode directly onto the membrane. The catalyst supported on carbon is placed directly on the surface of the electrolyte. It is thought that the catalyst can have a more efficient contact with the reactant gases and the catalytic yield can be improved (Valenzuela et al. 2009).

The assembly is formed by placing a gas diffusion layer, normally carbon cloth, onto the catalyzed membrane. Partial hot pressing is also applied to the assembly. With this method it is considered that the catalytic activity can be improved by forming a much more number of active sites where the redox reactions can occur. The electrical connection between the supported catalyst and the carbon cloth may be improved by the use of a nanocatalyst, but there is much discussion related to the influence of nanoparticles on the electro-osmotic drag mechanism of the polymeric membrane mass transport.

A well prepared assembly, independent of the method employed to make it, is part of the key to obtaining a fuel cell with excellent electrical characteristics.

Water management in electrodes is a problem that needs to be considered. Water can be formed as a by-product of the complete redox reactions in a PEMFC and also by the permanent hydration of the membrane to maintain its high value of proton conductivity. There is a problem with the amount of water flooding the electrolyte, which can cause a blocking effect in the gas diffusion layer zone. Water is necessary for the functioning of a PEMFC; humidification is part of the reactant gases or produced during the cathode reaction. In this case, it is necessary to design an appropriate fuel cell system (gases, operation temperature, and humidification) considering variables like relative humidity, water content, and vapor pressure. It is also important to mention that the humidification of the reactant gases plays an important role in the performance of the PEM fuel cell when it is operated at a temperature higher than 50 °C (Büchi and Srinivasan 1997).

The performance of PEMFC at any temperature is not 100 % efficient. The chemical to electrical energy conversion efficiency is up to 50 %. This means that the other 50 % or more is heat production. This heat can be useful power in a CHP or simply must be removed from the system. The Nafion membrane and MEA are considered to work at optimal conditions at temperatures below 100 °C. The heat produced by a fuel cell can be measured by analyzing the product water obtained from the fuel cell performance. Heat needs to be removed from the fuel cell and it is normal practice to use a fan or cooling with air or water, depending on the power capacity of the fuel cell. Cooling with air is proposed for fuel cells of up to 1 kW capacity, for larger power capacities water cooling is the most adequate system to remove heat from PEMFC.

The heat removed from a fuel cell can be released to the atmosphere or can be recovered and reused in domestic or industrial CHP systems; in this case, the mechanism to cool the fuel cell becomes much more practical and the application of this fuel cell becomes more attractive.

The experimental conditions for operating a PEMFC stack requires a rigid frame where the water produced and operating pressure of reactant gases need to be managed for safe operating conditions. It is also clear that research on bipolar plates is very significant in order to establish the conditions for operating a PEMFC. Bipolar plates are rigid structures placed on the anode and cathode of the fuel cell, two of their functions are the collection/delivery of electrons and injection of reactant gases to the MEA for carrying out the redox reactions in the fuel cell. Special attention is placed on bipolar plates is because about 80% of the mass of a PEMFC corresponds to the bipolar plates, affecting the effective energy density (Murphy *et al.* 1998). In the case of platinum, the amount of material utilized has been drastically reduced, but in the case of bipolar plates, the use of classical materials, for example graphite, continues to be a real problem that limits the energy density in a PEMFC.

Bipolar plates are still an unsolved problem, so the new technologies for designing and developing bipolar plates must be in accordance with the following considerations (Ruge and Büchi 2001):

1. The electrical conductivity must be higher than 10 S/cm.
2. The heat conductivity must be higher than 20 W/m*K for integrated cooling systems or must be higher than 100 W/m*K for just removing the heat from the plates.
3. The gas permeability must be less than 10^{-7} mbar L/s*cm^2.
4. The material used to make bipolar plates must be corrosion resistant when in contact with strong electrolyte, reactant gases, heat, and humidity.
5. The material also needs to be rigid, with a fracture strength higher than 25 MPa.
6. The bipolar plates must be slim, light, for maximizing the volume and energy density.
7. The cost of manufacturing bipolar plates must be as low as possible.
8. The channels of the bipolar plates carrying gases must be as functional as possible.

The rigid characteristics of PEM fuel cells allows performance at normal air pressure, as is observed in low power PEMFCs; nevertheless, when a PEM fuel cell operates at high power (10 kW or more), it is possible that the pressure requirements are higher. The advantages or disadvantages of operating fuel cells at high pressure are not completely understood; probably it is necessary when the performance of a PEMFC can be compared with an internal combustion engine, where the pressure is increased for increasing the specific power of the engine.

Hydrogen and oxygen internal pressures are associated to the covered fraction of gases reacting at the catalytic layer (MEA), and gas humidification is also considered to increase the output power density. PEMFC performance is a function of many factors needed to be considered when the fuel cell is operating as standalone or in a cogeneration system.

3.2.2 Direct Methanol Fuel Cells

The direct methanol fuel cell operates very similarly to PEMFC, but the use of methanol as the hydrogen source is the main difference. The fuel in DMFC (Figure 3.2.) is a liquid alcohol with higher storage efficiency (95 % efficiency) compared to hydrogen stored at 300 bar pressure in composite cylinders (0.6 % efficiency) or stored in metal hydride cylinders (0.65 % efficiency).

Methanol is an available and low-cost liquid fuel whose energy density is very similar to that of gasoline and can be stored and transported in containers that need not be safer than other containers with any conventional fuel (hydrocarbon). However, it is necessary to note that for human handling there are some security specifications to be considered.

There are many technological problems around DMFC, but one of the most important is the anodic oxidation reaction that occurs more slowly than hydrogen in PEMFC. Moreover, the methanol oxidation is a complex reaction that is not clearly understood. Another significant problem with DMFCs is the methanol crossover from the anode to cathode. The MEA structure is the same as that of PEMFC and it is strongly affected by the methanol/water absorption at the cathode. This process reduces significantly the I-V performance characteristics of the fuel cell. As a result of those problems, the DMFC operates with lower performance than other types of fuel cells, but is expected that DMFCs will be an important option in a near future for mobile and domestic applications.

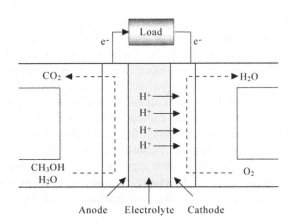

Figure 3.2 Schematic representation of the performance of a DMFC

Methanol Oxidation in a DMFC

The overall reaction in a DMFC is represented by

$$CH_3OH + 3/2O_2 \rightarrow 2H_2O + CO_2 \qquad (3.4)$$

This reaction has a change in molar Gibbs energy of $-698.2 \, kJ/mol$, and six electrons are transferred for each oxidized methanol molecule. The reversible cell voltage for this reaction is $E = 1.21 \, V$. In experimental DMFCs, the cell voltage is lower than $1.21 \, V$, indicating the losses present in this fuel cell (around $0.6 \, V$) due principally to anodic contribution where the alcohol oxidation reaction takes place.

Anodic Reactions in a DMFC

Direct methanol fuel cells are under research and the usual electrolyte is the Nafion membrane, so the functional performance of a DMFC is very similar to a solid electrolyte fuel cell. The overall anode electrochemical reaction can be written as

$$CH_3OH + H_2O \rightarrow 6H^+ + 6\,e^- + CO_2 \qquad (3.5)$$

The oxidation reaction of methanol at the anode of a DMFC implies the use of methanol and water to produce the half cell reaction. Protons (H^+ ions) move through the polymer electrolyte and the six electrons move across an external electrical circuit to complete the fuel cell reaction in the cathode (cathodic reaction) as

$$3/2 \, O_2 + 6H^+ + 6e^- \rightarrow 3H_2O \qquad (3.6)$$

At the end, the six electrons are transferred *per* methanol molecule in the complete redox reaction.

There are many mechanisms for oxidizing CH_3OH to CO_2 in a DMFC. One of the most accepted theories is the oxidation in steps, forming intermediates. In this case, the methanol molecule is firstly oxidized to formaldehyde, later to formic acid, and at the end the formation to carbon dioxide. Six electrons are transferred in the whole process. The concentration of methanol in the anode of a fuel cell is thought to be around 1 mol (Larminie 2003, Dohle *et al.* 2002). Using 1M CH_3OH it is possible to prevent or retard the methanol crossover to the cathode of the cell.

The catalytic materials to promote the oxidation reaction at the anode of the DMFC need to have a more sophisticate performance than the classical platinum used in PEMFC, since the methanol oxidation reaction proceeds more slowly than hydrogen oxidation, contributing to considerable activation over-potential at the anode as well as in the cathode of the DMFC. Much work has been carried out to develop an optimal catalyst for methanol oxidation in anodic reactions. Platinum is not a good proposal because there is a high possibility of CO absorption as a result of the intermediate reactions and species formed during the oxidation of CH_3OH to CO_2. Bimetallic catalysts have shown better results for oxidizing methanol in DMFCs (Hamnett and Kennedy 1988).

3.2.3 Alkaline Electrolyte Fuel Cells

The anodic electrochemical reaction occurring in an alkaline electrolyte fuel cell can be expressed as

$$2H_2 + 4OH^- \rightarrow 4H_2O + 4e^- \qquad (3.7)$$

The cathodic electrochemical reaction collects the electrons passing through an external circuit, forming OH^- *via* an electrochemical reduction reaction

$$O_2 + 4e^- + 2H_2O \rightarrow 4OH^- \qquad (3.8)$$

The electrolyte used by this type of fuel cell is an alkaline solution. Sodium hydroxide and potassium hydroxide solution are the mostly used electrolytes due to their low cost, high solubility, and stability at medium temperature. Neither hydroxide shows high corrosive properties. Potassium hydroxide has been widely used due to cost considerations. AFCs operate at different pressure, temperature, and electrode composition conditions according the designs. There are some examples of experimental AFCs operating at different temperatures from 90 to 260 °C. Apollo and Orbiter alkaline fuel cells are two good examples of extraordinary fuel cells. The sketch of an alkaline fuel cell is shown in Figure 3.3.

In the beginning of fuel cell research, PEMFCs took the advantage over AFC in NASA programs, but an alkaline fuel cell was used in Apollo missions to the moon. The results of using AFCs in space programs, as an important demonstration of showing the high power and versatility of AFCs, allowed the research of experimental fuel cells in the 1960s and early 1970s. Apart from space programs, in terrestrial applications AFCs were used in agriculture and power car projects (Ihrig 1960, Kordesh 1971, McDougall 1976, Kordesh and Simader 1996). It is possible that the advances in PEMFC in the last years has led to diminished interest in AFC research, but recently there have been many groups and companies in America and Europe, principally, working on new research on the possibility of

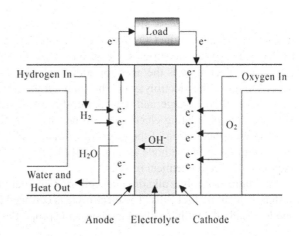

Figure 3.3 Schematic representation of an alkaline fuel cell

developing this type of fuel cell. AFC presents at least three advantages over other types of fuel cells. (1) The activation over-potential at the cathode of an AFC is lower than that of a fuel cell with acid electrolyte. This is a normal characteristic in all low temperature fuel cells, but the fundamental property of the oxygen reduction conditions to proceed faster in alkaline systems is not well understood. A normal value for the operating voltage *per* cell in an AFC is around 0.87 V and higher values are observed than in solid electrolyte fuel cells. (2) The fuel cell cost, in the case of the electrolyte is a great advantage over other types of fuel cells; potassium hydroxide is a cheap chemical reagent and very cheap with respect to polymer electrolytes. The cathode can be made from non-noble metals and this is another element that permits one to consider alkaline fuel cells as low cost fuel cells. The electrodes, in general, are considered to be of cheap materials with respect to other types of fuel cells. (3) AFCs normally do not use bipolar plates. This is a big advantage of this type of fuel cell because there is a significant reduction in the final cost of the fuel cell, but a disadvantage in this case is the low power density of AFC with respect to other fuel cells where bipolar plates are part of the system.

There are two kinds of alkaline fuel cells for the operating mode of the electrolyte: mobile and static electrolytes. The principal characteristics of the mobile electrolyte are established in the basic structure of the operating mode for injecting electrolyte to the fuel cell. HOH solution is pumped around the fuel cell and hydrogen is supplied to the anode. The excess of hydrogen and water produced in the anode need to be removed and cooled for continuous use. Normally, the hydrogen is supplied from a high pressure container and the water is condensed in a special part of the hydrogen recycling mechanism. Almost all AFCs perform with mobile electrolyte systems that allow refreshing of electrolyte, promoting the disposal of OH$^-$ ions at any moment. This is necessary because the carbon dioxide in the air reacts with the potassium hydroxide electrolyte as

$$2KOH + CO_2 \rightarrow K_2CO_3 + H_2O \qquad (3.9)$$

In this case, the potassium hydroxide is gradually changed to potassium carbonate and the OH$^-$ free ions are replaced by CO_3^{2-} ions, affecting the performance of AFC. This process always exists in an experimental cell because it is impossible to remove all the CO_2 from the air supply system; it is also recommended to change the carbonate enriched electrolyte at any time. The great disadvantage of the mobile electrolyte is the need for additional systems, for example, pumps are needed to move the electrolyte and the supplied fuel. This additional equipment is sensitive to fluid leakage and it incurs costs for the generated electrical energy. However, the circulating electrolyte can be used as a cooling system for the fuel cell and with the motion of the electrolyte it is possible to avoid the formation of solid solutions at the cathode as a result of the consumption of water according to the cathodic electrochemical reaction.

The static electrolyte alkaline fuel cell is an alternative to the mobile electrolyte system in alkaline fuel cells. The electrolyte is confined in a porous container from anode to cathode, with excellent anticorrosion properties. This kind of AFC neces-

Table 3.1 Operating conditions for the most representative AFCs (Larminie and Dicks 2003)

Fuel cell	Pressure bar	Temperature °C	KOH (% conc.)	Anode catalyst	Cathode catalyst
Bacon	45	200	30	Ni	NiO
Apollo	3.4	230	75	Ni	NiO
Orbiter	4.1	93	35	Pt/Pd	Au/Pt
Siemens	2.2	80	NA	Ni	Ag

sarily uses pure O_2 at the cathode to avoid electrolyte contamination. Static electrolyte alkaline fuel cells are sealed, and it is not possible to change the electrolyte in case of carbonate formation at the electrolyte. The hydrogen is circulated as in the case of mobile electrolytes to remove water formation at the anode. The water can be collected for other applications. It is necessary to implement a cooling system due to the operating conditions of the fuel cell. The cooling fluid can be water or more sophisticated cooling fluids like glycol/water or fluorinated hydrocarbon dielectric liquid (Warshay and Prokopius 1990). The implementation of AFC performing in mobile or static mode depends on the specific application; probably it is easier to remove the electrolyte for maintaining a proper operation of the fuel cell.

AFCs operate at higher ambient pressure and temperature. The main operating parameters in the most representative AFC are given in Table 3.1. The operating conditions of AFC are correlated together (temperature, pressure, KOH concentration, and catalysts).

From Table 3.1 it can be seen that the temperature, pressure, and electrolyte concentration for every AFC has been adjusted to obtain the best operating conditions. It is not possible to establish a criterion to indicate the optimal values in any example mentioned in Table 3.1 and, obviously, the operating conditions have been proposed according to the final space or terrestrial application.

It is common that high pressure containers are used in alkaline fuel cells and sometimes this adds high costs for using this pressured system. In addition to gas containers, gas leakage and internal stresses of reactants are variables that must be controlled.

3.2.3.1 Common Electrodes Used in Alkaline Fuel Cells

AFCs can be operated at a wide range of temperatures and pressures but their range of applications is not as extensive as that of PEMFCs. There are no typical or standard electrodes for the AFC, and depending of the specific applications, the performance requirements, budget, operating temperature, and pressure, different electrodes have been investigated. Different catalysts have also been investigated but are independent of the electrode structure. Platinum is probably the most used catalyst in this type of fuel cell.

There are three main electrode structures used in AFCs: (1) sintered nickel powder, where porous electrodes are made by sintering nickel powder to obtain a rigid structure. (2) Raney metals are used to obtain very active and porous electrodes. The electrodes are prepared by mixing the catalytic material (probably nickel) with an inactive metal (usually aluminum), and the mixture is treated in strong alkali to dissolve the aluminum forming a porous material with very high surface area. Due to the preparation conditions, this type of electrode normally gives good results in AFCs. (3) Rolled electrodes have been developed by using carbon supported catalysts, and in this case, the electrode preparation implies the use of PTFE. The supported catalyst and PTFE are mixed together and placed on an electrical conducting material such as nickel mesh or foam. Some problems with this type of electrode can be the non-conductive area due to the presence of PTFE.

3.2.3.2 Technological Challenges for the Development of Alkaline Fuel Cells

The carbon dioxide reaction with the electrolyte in AFCs is the main problem that needs to be solved. CO_2 may also be present in the hydrogen as fuel, if the hydrogen is derived from hydrocarbons. In general, the CO_2 reaction in an AFC affects the performance of the fuel cell, and the effects of this reaction are principally the formation of carbonate compounds and the reduction in OH^- ion concentration in the anode side of the fuel cell (Larminie and Dicks 2003). The most interesting performances of AFCs have been obtained by using pure hydrogen and oxygen, so the new research direction needs to be focused on developing optimal AFC components and also the adequate way to properly use the generated heat from the performance of this type of fuel cell.

3.2.4 Phosphoric Acid Fuel Cells

The phosphoric acid fuel cell was the first fuel cell utilized in terrestrial applications; phosphoric acid may be used in concentrated ratios (above 85 %) and at intermediate temperatures (150–200°C).

In this type of fuel cell, the acid is used as the ionic electrolyte as well as the polymeric membranes in PEMFC. The chemical reaction of the acid is

$$H_3PO_4 \rightarrow H^+ + H_2PO_4^- \tag{3.10}$$

And the solvation process occurs as

$$H^+ + H_3PO_4 \rightarrow H_4PO_4^- \tag{3.11}$$

The chemical specie H_3PO_4 is the real solvent and responsible for the conducting mechanism (Grotthus type).

When a PAFC operates at intermediate temperature (about 150 °C), the acid concentration needs to be about 85 %. In other experiments the proposed concen-

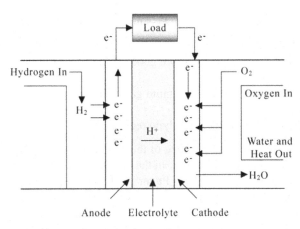

Figure 3.4 Performance of a PAFC

tration has been almost 100 % for higher operating temperatures, but the performance has not been satisfactory. The acid reacts at 150 °C as

$$2 H_3PO_4 \rightarrow H_4P_2O_7 + H_2O \tag{3.12}$$

In this case, $H_4P_2O_7$ has a higher ionization constant than H_3PO_4. Then it is possible that

$$H_4P_2O_7 \rightarrow H^+ + H_3P_2O_7 \tag{3.13}$$

At this temperature or higher, the activation over-potential losses decrease, especially at the cathode side, because $H_4P_2O_7$ is adsorbed weakly compared to phosphoric acid anion on the catalyst (in this case platinum is considered) and the conductivity of $H_4P_2O_7$ is also higher then phosphoric acid.

The performance of a PAFC is very similar to PEMFC. PAFC utilizes a liquid proton conducting electrolyte. The oxidation reaction occurs at the anode and the reduction reaction at the cathode of the fuel cell (Figure 3.4). In the PAFC, the electrochemical reactions take place on highly dispersed electrocatalyst particles supported on carbon black. The electrolyte, a concentrated phosphoric acid, like the membrane in the PEM cells, is used to transport ions.

Phosphoric acid is stable (thermal, chemical, and electrochemical) at temperatures above 150°C and is the unique liquid acid considered as an electrolyte for fuel cells. Phosphoric acid shows excellent tolerance to CO_2 as part of the fuel or the oxidant.

3.2.4.1 Electrode Components and the Catalyst

The PAFC uses gas diffusion electrodes. PTFE/Pt-black in porous electrodes with loadings of 9 mg Pt/cm^2 are used in both electrodes (anode and cathode). Other catalytic electrodes have been obtained by using Pt supported on carbon. The carbon has the following important functions:

1. It promotes the dispersion of the catalyst to ensure good utilization of the catalytic material.
2. It provides micropores in the electrode for maximum gas diffusion in the electrode/electrolyte interface.
3. It increases the electrical conductivity in the catalytic region.

When the catalyst is well dispersed in the surface of the carbon, the amount of platinum loading has been reduced in the electrodes. In the last two decades, the Pt loading was significantly decreased, about $0.10\,mg\,Pt/cm^2$ in the anode and about $0.50\,mg\,Pt/cm^2$ in the cathode. The activity of the Pt catalyst depends on the nature of catalyst, particle size, and specific surface area. Small crystallites, an adequate dispersion, and high surface area generally lead to high catalyst activity. Actually it is possible to find nanostructured Pt loadings, with particle size distribution around 2–4 nm and a high surface area of up to $100\,m^2/g$. It is expected that the performance of the fuel cell may be improved by using nano-catalysts supported on carbon (Appleby 1984, Kordesch 1979, Kinoshita 1988). However, electrode performance does decay with time. This is due primarily to the agglomeration of Pt catalyst particles and the obstruction of gases through the porous structure caused principally by electrolyte flooding. Carbon has shown corrosion problems at high cell voltages (above ~0.8 V).

3.2.4.2 Characteristics of Phospohric Acid Fuel Cell Operations

Research on PAFCs was conducted quite extensively from about 1975 to 1985 because the PAFC system was projected to be the first of all types of fuel cells to find an application as a stationary power source. Indeed, the 200 kW PAFC power plant (PC 25 manufactured by UTC-International Fuel Cells) was designed as an on-site integrated energy system (CHP) in 1996. The research and development of PAFC have shown the advantages of this technology, but there is strong competition between low and intermediate-temperature fuel cells for terrestrial applications in mobile or stationary projects.

The performance of PAFCs in single cells was very low compared to that of polymer electrolyte or alkaline electrolyte fuel cells at operating temperatures below 100 °C. Strong adsorption of the phosphate ion on the Pt catalyst led to high over-potentials in concentrated phosphoric acid fuel cells. To solve this problem, the temperature was adjusted to 150°C and the electrolyte concentration was maintained at 85 %. There were considerable reductions in activation and ohmic over-potentials at this temperature. As has been mentioned before, at this temperature, phosphoric acid polymerizes to pyrophosphoric acid, improving its characteristics as a strong acid with a high ionic conductivity. Also, the pyrophosphate anion is considerably larger than the phosphate anion, reducing its specific adsorption considerably. In further studies, the performance of the fuel cell was investigated at more concentrated acidic media (95 and 10%). The optimum operating temperature of this type of fuel cell is around 200°C. The benefits of such an operating temperature are better electrode kinetics of oxygen reduction, a decrease in the ohmic over-potential losses, and a higher CO tolerance.

There is an extensive research related to the determination of the corrosion resistance and corrosion rates of a variety of carbons that have been used as catalytic supports for efficient and high active area catalysts (Pt and Pt compounds principally). Examples of carbon supports are Vulcan 72, acetylene black, *etc.* Some supports have a lower corrosion rate, with a small active area. In general, it is possible to establish that the corrosion rate increases with increasing electrode potential and is high at the open circuit potential.

The use of catalytic nanoparticles is a problem in PAFC due to the extreme operating conditions. Dissolution of nanoparticles has been observed in experimental fuel cells performing at nominal conditions. Sometimes it forms a precipitate of supported catalyst, and in other cases, the operating conditions of a PAFC lead to redeposit on the larger Pt particle, causing a loss of active surface area in the catalytic region. The problem is more severe when the PAFC operates close to the open circuit potential.

There is a smaller effect of CO poisoning on the over-potential at the hydrogen electrode in PAFCs than that in the case of low temperature acid electrolyte fuel cells. The adsorption of hydrogen increases at higher temperatures and CO adsorption decreases significantly for a better performance of the PAFC. Other significant poisoning to the catalyst is hydrogen sulfite, which can be formed simply by reacting impurities, so the operation of a PAFC has some implications that need to be solved for a more simple and standard use in terrestrial applications.

Extensive investigations have been carried out on PAFC single cells and cell stacks to determine their lifetime characteristics. The target for a fuel cell stack has been a continuous lifetime of 40,000 h with a performance degradation of 1 mV/1000 h for stationary applications. The degradation rate was about 10 times the target value. Several factors contributed to the degradation, including corrosion of the carbon support and Pt dissolution or precipitation. Performance degradation was also caused by poisoning of the Pt anode electrocatalyst by CO and H_2S. In the last 30 years there has been a significant progress in the development of more efficient PAFCs, and research has been focused on identifying the causes for the degradation in lifetime. Current PAFCs operate with a cell potential at 0.8 V or lower and try to minimize the corrosion rate of the carbon support and platinum particles or any other catalyst. An alternate method is to have the cathode in an inert atmosphere when the cell is in open-circuit conditions. The fuel is CO-free or is minimum in the system. However, if reformed hydrogen is used, the CO_2 content in this gas is reduced to CO and raises the concentration above the tolerance level. The performance degradation can also be reduced by the complete elimination of sulfur in natural gas by hydro-desulfurization (Srinivasan 2006).

The PAFC stack consists of a repeating arrangement of a bipolar plate, the anode, electrolyte matrix, and cathode. In a manner similar to that described for the PEM cell, the bipolar plate is used to separate the individual cells and permit the interconnection in series. At the same time, it provides reactant gases in the anode and the cathode. There are many designs for the bipolar plate and stack components to obtain maximum power, stability, and lifetime from the PAFC (Appleby and Foulkes 1993).

In PAFC stacks, as part of the fuel cell design, a system to remove the heat generated during the operation is considered. Liquid (water/steam or a dielectric fluid) or gas (air) coolants that are routed through cooling channels or pipes located in the cell stack have been proposed for an efficient management of heat removal. Liquid cooling requires complex manifolds and connections, but better heat removal is achieved with air cooling. The advantage of gas cooling is its simplicity, reliability, and relatively low cost. However, the size of the cell is limited, and the air-cooling channels are much larger than those needed for liquid cooling. Water cooling is the most popular method and has been used in demonstration systems.

Water cooling can be used with boiling water or pressurized water. Boiling water cooling uses the latent heat of vaporization of water to remove the heat from the cells. Since the nominal operating temperature of the fuel cell is around 180–200 °C, the temperature of the cooling water can be up to 150–180 °C, which increases the efficiency of the cell.

Pressurized water cooling is also used as the cooling system, but in this case, the heat is only removed from the stack by the heat capacity of the cooling water, so the cooling is not as efficient as with boiling water. The main disadvantage of water cooling is that the purification system must be coupled to prevent corrosion of cooling pipes and connectors. The water quality required is similar to that used for boiler feed water in conventional thermal power stations.

3.2.4.3 Experimental Phosphoric Acid Fuel Cell Systems

PAFCs with a high performance and feasibility are commercially available. This type of fuel cell is also under development but it has some characteristics that are considered as the most developed in comparison to the other types of fuel cells discussed in this chapter. As an example of this development there are fuel cell power systems available in the market with normalized specifications. Many of these systems have now run for several years in demonstrative projects and it has been possible to learn and improve the technology for this type of fuel cell (Fuel Cells Bulletin 2008).

The new PAFC systems have been considered as leading systems for supplying energy with a new technological approach and there are many demonstrative projects in banks, hospitals, and in general in stationary applications where the opportunity area is also in CPH systems.

References

Appleby J (1984) In: Sarangapani S, Akridge JR, Schumm B (eds) Proceedings of the Workshop on the Electrochemistry of Carbon, The Electrochemical Society, Pennington, p 251

Appleby AJ, Foulkes FR (1993) A fuel cell handbook, 2nd ed. Krieger, New York

Büchi FN, Srinivasan S (1997) Operating proton exchange membrane fuel cells without external humidification of the reactant gases. Fundamental aspects. J Electrochem Soc 144(8): 2767–2772

Dohle H, Schmitz H, Bewer T, Mergel J, Stolten D (2002) Development of a compact 500 W class direct methanol fuel cell stack. J Power Sources 106:313–322

EG&G Services Parsons, Inc. (2000) Fuel Cell Handbook, 5th edn. Science Applications International Corporation. National Energy Technology Laboratory

Fuel Cells Bulletin (2008) UTC Power to launch improved, cheaper PAFC, 5:6–7

Hamnett A, Kennedy BJ (1988) Bimetallic carbon supported anodes for the direct methanol air fuel cell. Electrochimica Acta 33:1613–1618

Ihrig HK (1960) 11th Annual Earthmoving Industry Conference, SAE Paper No. S-253

Kinoshita K (1988) Carbon: electrochemical and physicochemical properties, Wiley Interscience, New York

Koppel T (1999) Powering the future – the Ballard fuel cell and the race to change the world. Wiley, Toronto

Kordesh KV (1971) Hydrogen-air/lead battery hybrid system for vehicle propulsion. J Electrochem Soc 118(5):812–817

Kordesch KV (1979) Survey of carbon and its role in phosphoric acid fuel cells. BNL 51418, prepared for Brookhaven National Laboratory

Kordesh KV, Simader G (1996) Fuel cells and their applications. VCH Verlagsgesellschaft, Weinheim

Larminie J, Dicks A (2003) Fuel cells systems explained, 2nd ed. UK

McDougall A (1976) Fuel cells. Macmillan, London

Murphy OJ, Cisar A, Clarke E (1998) Low cost light weight high power density PEM fuel cell stack. Electrochimica Acta 43(24):3829–3840

Rozière J, Jones D (2001) Recent progress in membranes for medium temperature fuel cell. In: Proceedings of the First European PEFC Forum (EFCF), pp 145–150

Ruge M, Büchi FN (2001) Bipolar elements for PEM fuel cell stacks based on the mould to size process of carbon polymer mixtures. In: Proceedings of the First European PEFC Forum (EFCF), pp 299–308

Srinivasan S (2006) Fuel cells. Springer, New York

Valenzuela E, Gamboa SA, Sebastian PJ, Moreira J, Pantoja J, Ibañez G, Reyes A, Campillo B, Serna S (2009) Proton charge transport in Nafion nanochannels. J Nano Res 5:31–36

Warshay M, Prokopius PR (1990) The fuel cell in space: yesterday, today and tomorrow. J Power Sources 29:193–200

Zhang YY, Quing-Chun Y, Cao GY, Zhu XJ (2006) Research on a simulated 60 kW PEMFC cogeneration system for domestic application. J Zhejiang University Science A 7(3):450–457

Chapter 4
State of the Art of Sorption Refrigeration Systems

4.1 Introduction

Sorption cooling systems have been used commercially for some decades for different applications including air conditioning and refrigeration, using a diverse range of thermodynamic cycles and technologies for many size and capacities. However, their use has been limited mainly because of their low efficiency and high investment costs, at least compared with compression systems that are widely used all over the world. Because of this, sorption and desiccant systems have been used, in general, only when large amounts of waste thermal energy that can be used as the energy supplied to the system are available, and recently with, for example, solar and geothermal technologies.

As will be explained in the next chapter, desiccant cooling (DEC) and sorption systems are in fact heat pumps since they have the capacity to absorb heat from a source at low temperature and to pump it to a heat sink at a higher temperature level. Depending on the use, common sorption systems are classified as sorption refrigeration systems when they are used for refrigeration and air conditioning, heat pumps when they are used for heating and heat transformers when they are used also for heating but the temperature of the useful heat is higher than the temperature of the heat supplied to the system.

Sorption systems can be classified depending on the sorption mechanism employed to pump the heat through the system to produce the cold. These mechanisms are absorption and adsorption and are explained in Chapter 6.

In this chapter, the state of the art of sorption refrigeration systems including absorption and adsorption systems are presented. The state of the art includes commercial systems, under development systems, and research studies including advanced systems and new working fluid mixtures. All this, is studied from the point of view of the sorption systems, which have real potential for use in co-generation systems together with PEM fuel cells. This means that only those systems operating at temperatures lower than 90 °C, which are in general small capacity systems, will be considered in this chapter.

4.2 Commercial Systems

Since about 50 years ago, there have been some major companies offering sorption systems in the range of high capacities from 200 kW up to few MW in the market. Most of these systems are direct gas fired, but there are others that are driven by steam or are oil fired. Most of the companies have developed absorption single-effect and double-effect units using a water–LiBr mixture, but there are others operating with ammonia/water. Some of the world's leading companies developing high capacity systems are: Carrier, York, Trane, Robur, Broad, Mycom, LG, Mitsubishi, Sanyo, Mc Quay, Entropie, Century, and Colibri. Because of the high capacity of these chillers, most of them have been used in large buildings and in industrial processes.

Whereas a considerable number of companies developed chillers of high capacity many years ago, just a few have developed systems of medium and small capacities. However, because of the considerable increases in the demand for air conditioning in the residential sector, nowadays there is a significant number of companies offering sorption systems of small capacities (<100 kW) utilizing diverse technologies and working mixtures.

Some of the most important companies that have developed their own machines for small capacities are: Nishyodo, Maekawa, Maycom, Yazaki, Robur, Broad, Rotartica, Climatewell, Sor Tech, Invesnsor, Thermax, Solar Next, Aosol, Pink, Sonnenklima, EAW, among others.

There are no reliable data in the literature about the number of thermally driven systems for cooling of small capacity installed worldwide, however, it was estimated that in 2007 there were between 250 and 300 cooling systems driven with solar energy operating in the world. About 200 of these systems have been installed in Europe and the most of these in Germany and Spain (Meyer 2008).

In an International Energy Agency overview of solar cooling systems for commercial buildings, it was found that from 88 large scale solar cooling plants, approximately 70 % were absorption systems, 13 % adsorption systems, and 17 % DEC systems operating with both solid and liquid desiccants, as can be seen in Figure 4.1 (Troi *et al.* 2008). From this data, it is clear that absorption cooling technology is the most predominant in the market, at least driven with solar energy.

In a study developed by the International Energy Agency in which data were obtained from 280 installed cooling machines, the result showed that 43 % were manufactured by Rotartica, 35 % by Climatwell, 7 % by EAW, 5 % by Yasaky, and the rest by various companies (Mugnier *et al.* 2008). Figure 4.2 shows the distribution of cooling system manufacturers.

In Sections 4.2.1 and 4.2.2 a brief description of some of the main manufactures of absorption and adsorption systems are presented, respectively. Section 4.2.3 shows and compares absorption and adsorption machines of small capacities whose characteristics have been found in the open literature or on web sites.

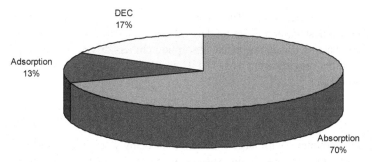

Figure 4.1 Distribution of cooling technologies in the market

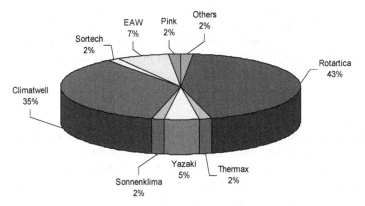

Figure 4.2 Distribution of cooling system manufacturers

4.2.1 Absorption Chillers

Yazaky

Yazaky is the world's leading Japanese company selling absorption chillers of small capacities (up to 105 kW). The company has developed absorption single-effect units driven by hot water and double-effect direct gas fired units, both operating with the water–LiBr mixture. The models driven by hot water are WFC-SC(H)10, 20, and 30 with cooling capacities of 35, 70, and 105 kW, respectively. These models have been commercialized for many years and recently the company launched into the market the model WFC-SC5 with a cooling capacity of 17.6 kW to compete with new developments of small capacities. For all the models, the nominal hot water temperature is 88 °C to produce chilled water up to 7 °C with a COP of 0.7.

Broad

Nowadays the Chinese Broad company is the world's largest manufacturer of air conditioners. Broad absorption chillers use a wide range of energy sources, such as oil, gas, steam, solar energy, *etc*. The capacities varied from just a few kW to some MW. The smallest units are the models BJ with capacities of 16, 23, and 64 kW. These are water–LiBr double effect absorption chillers driven by hot oil or steam at temperatures ranging from 165–180 °C assisted by a natural gas burner with COP of about 1.2.

Robur

Robur is an Italian company with large experience for many years in the development of heat pumps and air conditioning systems using direct gas fired absorption technology for residential and industrial applications. Recently, it modified the Robur GAHO-W model, placed a generator heat exchanger instead of a direct gas fired generator to be heated by a heat transfer fluid. The developed chiller has a capacity of 26 kW and operates with an ammonia–water mixture at temperatures of around 200 °C. Two prototypes have been installed in Italy and they have been in operation since 2006 (Häberle *et al.* 2007).

Rotartica S.A.

Recently the Spanish company Rotartica S.A., developed two small single effect absorption chillers operating with water–LiBr of 4.5 kW of cooling capacity. The absorption process takes place within a rotary chamber that increases the heat and mass transfer mechanisms, which makes the systems more efficient and compact. The model Solar 045 V rejects the heat to the atmospheric air without using cooling towers, which is an advantage not only for capital cost reduction but also for maintenance costs. The model Solar 045 was developed for applications where cooling towers are required. These systems operate at temperatures between 85 and 90 °C with a nominal COP close to 0.7.

ClimateWell A. B.

The Spanish-Swedish company developed a chiller unit denominated CW10, which has the ability to operate in three modes, charging, heating, and cooling. In the charging mode the device stores energy through drying the LiCl salt that subsequently can be used whenever required. This machine operates according to the principle of a thermo-chemical heat pump in a cyclical process between three states of aggregation solid, liquid, and gas. This device can charge and discharge simultaneously, which means that it can always receive thermal energy allowing continuous cooling and heating output. The CW10 unit consists of two parallel storage cells, each one containing a reactor and a condenser/evaporator that allows conversion of the intermittent operation cycle to a continuous one. The maximum capacity is 10 kW if only one accumulator is used for cooling or 20 kW if both accumulators are used simultaneously. This system operates at temperatures higher than 83 °C with thermal efficiencies of 68 and 85 % for cooling and heating, respectively.

EAW Energieanlagenbau Westenfeld GmbH

The company EAW and the Institute of Air Conditioning and Refrigeration both from Germany have developed some absorption air conditioning systems of different capacities operating with the water–LiBr mixture. Two of these systems are actually commercialized by the German company Schüco. The models LB 15 and LB 30 have an output cooling capacity of 15 and 30 kW, respectively. Because of their size, these systems are suitable for offices, schools, or hotels. They are optimized to be driven with water temperatures from 80–90 °C, producing cooling water at temperatures as low as 11 °C with a COP of 0.75.

SK Sonnenklima GmbH

The Phönix system, manufactured and commercialized by the company SK SonnenKlima GmbH, was developed by the German companies Phönix Sonnen Wärme AG and Zae Bayern together with the Technische Universität Berlin. The Phönix chiller is a single-effect cooling system operating with water–LiBr. This system was designed to obtain high COP at partial load and overload capacities. This system can operate at temperatures as low as 65 °C. The nominal capacity is 10 kW and the maximum COP is 0.77 at hot water temperatures between 80 and 90 °C.

Pink GmbH and Solar Next AG

The PSC 12 system was developed by the companies Solar Next AG from Germany, Pink GmbH from Austria, and the Joanneum Research Institute. This is an absorption single-effect system operating with an ammonia–water mixture. One of the main differences of this system with respect to commercial air conditioning systems is that it operates not only as air conditioning but also as refrigeration because it operates with ammonia as the refrigerant. Some of the peculiarities of this system are the novel membrane solution pump and the vertical falling-film tubular absorber. The novel Chillii unit has a cooling capacity of 10 kW and compact design. The cooling water temperatures reached were from +15–(−5) °C at heating temperatures from 65–115 °C. The maximum coefficient of performance is 0.64 (Jakob and Pink 2007).

4.2.2 Adsorption Chillers

Nishiyodo and Mayekawa are two major Japanese manufactures that have been developed adsorption chillers within the range of small and high capacities for some time. The German company GBU GmbH (Nishiyodo seller for Europe) commercializes a 70 kW chiller operating with water–silicagel, while Mycom (Mayekawa seller for Europe) commercializes a 50 kW chiller using the water–zeolith pair. These are the smallest small capacity systems commercialized by these companies.

In the last years Fraunhofer ISE and SorTech AG, both from Germany, have developed adsorption chillers for small scale applications. The models ACS 08 and ACS 15 with nominal cooling capacities of 7.5 and 15 kW, respectively, operate with the water–silicagel pair. Both units operate at temperatures as low as 60 °C (up to 95 °C). Single modules can be interconnected to achieve a cooling capacity as high as 60 kW. The nominal COP is 0.56 and the chilled water is in the range 6–20 °C.

4.2.3 Absorption and Adsorption Small Capacity Systems

Table 4.1 compares the small capacity absorption and adsorption chillers available in the market that are suitable to be coupled to PEM fuel cells since they can operate at temperatures of around 80 °C or lower.

Table 4.1 Absorption and adsorption small capacity systems available in the market

Company	Yazaki	Sonnenklima	EAW	Rotartica
Product name	WFC SC5, SC10, SC20, SC 30	Suninverse 10	Wegracal SE 15, SE 30, SE 50	Solar 045v, 045
Picture	Source: Yazaky	Source: Sonnenklima	Source: EAW	Source: Rotartica
Technology	Absorption	Absorption	Absorption	Absorption
Working pair	Water–LiBr	Water–LiBr	Water–LiBr	Water–LiBr
Nominal cooling Capacity [kW]	17.5, 35, 70, 105	10	15, 30, 54	4.5
COP [dimensionless]	0.72	0.78	0.71, 0.75, 0.75	0.7
Heating temperature in/out [°C]	88/83	75/ –	90/80, 90/80, 86/71	90/85
Chilled water in/out [°C]	12.5/7.0	18/15	17/11, 17/11, 15/9	13/10
Cooling water in/out [°C]	31/35	27/35	30/36, 30/35, 27/32	30/35
Dimensions (L×W×H) [m×m×m]	0.6×0.8×1.94, 0.76×0.97×1.92, 1.06×1.30×2.03, 1.38×1.55×2.07	1.08×0.69×1.90	1.75×0.76×1.75, 2.14×0.97×2.2, 2.95×1.1×2.31	1.09×0.76×1.15, 1.05×0.68×0.85
Weight [kg]	420, 604, 1156, 1801	550	660, 1400, 2250	290, 240

Table 4.1 Continued

Company	Climatewell	SolarNext	SorTech AG	SJTU
Product name	Climatewell 10	Chillii PSC12	ACS 08, ACS 15	SWAC-10
Picture				
	Source: Climatwell	Source: SolarNext	Source: SorTech	Source: SJTU
Technology	Absorption	Absorption	Adsorption	Adsorption
Working pair	Water–LiCl	Ammonia–water	Water–silica gel	Water–silica gel
Nominal cooling capacity [kW]	10	12	7.5, 15	10
COP [dimensioless]	0.68	0.62	0.56	0.39
Heating temperature in/out [°C]	73/ –	85/78	72/65, 72/66	85/79
Chilled water in/out [°C]	12.5/7.0	12/6	18/15	15/10
Cooling water in/out [°C]	– /17	24/29	27/32	30/36
Dimensions (L×W×H) [m×m×m]	$1.20 \times 0.80 \times 1.60$	$0.8 \times 0.6 \times 2.2$	$0.79 \times 1.06 \times 0.94$, $0.79 \times 1.35 \times 1.45$	$1.80 * 1.20 * 1.40$
Weight [kg]	875	350	260, 510	1600

4.3 Systems under Development

Some of the systems that are not yet commercialized but because of their high development degree but are expected to be introduced into the market in a short period of time are presented in this section. Experimental and theoretical studies on the development of sorption systems are presented in Section 4.4.

Solar Next AG (Germany) and Pink GmbH (Austria) have developed a single-effect ammonia–water absorption chiller denominated chillii® PSC with nominal cooling capacities of 5, 10, and 20 kW (Jacob and Pink, 2007). Because of the use of ammonia as refrigerant, these systems can be used not only for air conditioning but also for refrigeration. The design characteristics, the coefficient of performance, and the operating temperature are similar to those obtained with the already commercialized chiller PSC12 described in the previous section.

The Institute of Thermodynamics and Thermal Engineering (ITW) of the University of Stuttgart has developed a 10 kW air cooled absorption chiller operating with the ammonia–water mixture. All the components are built as plate heat exchangers with the exception of the generator. The system has been tested at heat supplied temperatures from 82–117 °C to produce chilled water in the range 5–16 °C and the COP has varied from 0.58–0.74 (Zetzsche *et al.* 2007).

The well known Robur company has developed a small ammonia–water absorption machine of 16.9 kW cooling capacity to produce chilled water at 7 °C at nominal capacity (Mugnier 2008).

The Austrian Company Solar Frost has developed an ammonia–water chiller with a variable range of 2–10 kW (2 kW *per* module) cooling capacity. The system has been designed to operate with hot water at temperatures between 70–120 °C to produce cooled water at temperatures as low as 3 °C. (Mugnier 2008).

The solar collector manufacturer Aosol in Portugal is developing an absorption air cooler with 6 kW cooling capacity operating with the ammonia–water mixture (Mugnier *et al.* 2008).

The Institute of Thermal Engineering at Graz University of Technology has developed a small-capacity single-stage ammonia–water absorption heat pump with 5 kW cooling capacity. The unit has been designed to operate with cold water temperatures ranging from −10–20 °C and cooling water temperatures ranging from 30–50 °C. The COP varies from 0.75–0.4 depending of the temperature lift (Moser and Rieberer 2007).

4.4 Research Studies

Some of the most important experimental and theoretical studies on the development of sorption cooling systems are presented in Sections 4.4.1 and 4.4.2, respectively. Each section has been also divided in advanced systems, alternative mixtures, and components.

4.4.1 Experimental Studies

4.4.1.1 Advanced Systems

Experimental studies for the development of advanced systems have been proposed by various authors in the last years. The research was carried out in order to find new configurations with higher coefficients of performance, temperature lifts, or lower heat source temperatures. Also, research has been carried out on cogeneration systems in which the heat dissipated for one of the systems is used as heat input in the sorption units. Some of the most relevant recent investigations are the following.

Yaxiu *et al.* (2008) did experimental research on a new solar pump-free lithium bromide absorption refrigeration system with a second generator. The results showed that with the proposed system the required minimum driving temperature of the heat source was only 68 °C compared with about 100 °C required in traditional absorption refrigeration systems. The maximum coefficient of performance obtained from the evaluation approached 0.8.

Caeiro (2008) reported experimental results of an innovative hybrid refrigeration cycle bringing together a single-effect water–lithium bromide refrigerator and a steam ejector and aimed at vacuum freezing and ice storing. The average coefficient of performance of the freezing process was 0.24.

Wagner *et al.* (2008) developed and tested a waste heat-driven chiller capable of simultaneously generating chilled and heated water. In addition to simultaneous operation, the system has the ability to seamlessly change from chilling to heating eliminating the seasonal changeover which is typical in existing absorbers. These enhancements significantly increase the flexibility and utilization of the chiller.

At the Energy Research Center of the National University of Mexico an absorption refrigeration system of small capacity between 5–10 kW operating with an ammonia–lithium nitrate mixture is being developed. The condenser and absorber are both air cooled and the heat is supplied by evacuated tube collectors. Due to the use of ammonia as refrigerant, the prototype can operate as a refrigeration or an air conditioning system. The design COP is 0.53 for heat supplied at 120 °C, rejected heat at 40 °C, producing chilled water at 0 °C (Llamas *et al.* 2007).

Gómez *et al.* (2007) reported a theoretical and experimental evaluation of an indirect-fired GAX cooling system using ammonia–water as the working fluid with a cooling capacity of 10.6 kW. The system operates at temperatures ranging from 180–195 °C and it is air cooled. The authors concluded that the system presents potential to compete technically in the Mexican air conditioning market.

At the Ecole Nationale d'Ingénieurs de Monastir in Tunisia an air cooled absorption-diffusion machine working with hydrocarbons as refrigerant/absorbent (n-butane/n-octane) pair and helium as pressure equalizing gas with a refrigerant capacity of 100 W is being developed. The evaporator temperature is 6 °C with driving temperature of 125 °C (Ezzine *et al.* 2007).

At the Technical University of Ilmenau in Germany an absorption refrigeration machine for cooling and air conditioning with a cooling capacity of 10 kW is benig evaluated. The machine has been tested with a new proposed acetone–zinc bromide mixture. The results have showed that the absorption machine can operate with drive heating source temperatures as low as 55 °C but with low coefficients of performance (Ajib and Gunther 2007).

Sun and Guo, (2006) reported the results of a prototype of combined vapor compression–absorption refrigeration system where a gas engine directly drives an open screw compressor in a vapor compression refrigeration chiller and waste heat from the gas engine is used to operate the absorption refrigeration cycle. The cooling capacity of the prototype reached was about 589 kW, producing chilled water at 7 °C. The calculated results showed that the primary energy rate of the prototype was about 1.81, saving more than 25 % of primary energy compared to a conventional electrically driven vapor compression refrigeration unit.

Arivazhagan *et al.* (2006) investigated the performance of a two-stage half-effect vapor absorption cooling system of 1 kW of capacity using HFC based working fluids (R134a as refrigerant and DMAC as absorbent). The system is capable of producing evaporating temperature as low as −7 °C with generator temperatures ranging from 55–75 °C. The optimum generator temperature was found to be in the range of 65–70 °C with coefficient of performance of 0.36.

Glebov and Setterwall (2003) studied the influence of 2-methyl-1-pentanol (2 MP) on the cooling effect of pilot absorption chiller. The experimental results showed that the presence of the additive in the vapor phase, favors the cooling enhancement more than the additive in the LiBr solution. The enhancement heat transfer ratio was between 20 and 30 %. Also, it was noticed that the additive travels around the absorption cycle during long-term operation.

4.4.1.2 Alternative Mixtures

Experimental studies have been carried out in order to find alternative mixtures to the water–lithium bromide and ammonia–water mixtures, which have been used the most for air conditioning and refrigeration, respectively. Some of the most relevant studies of new mixtures that have been tested in complete absorption cooling systems were presented in the previous section. Other alternative mixtures that have been tested only in some components such as absorbers, generators, *etc.*, are presented in the present section.

Muthu *et al.* (2008) reported the experimental results of a vapour absorption refrigeration system operating with the R134a–DMAC mixture, the results indicated that at the sink and source temperatures of 30 and 80 °C respectively, the coefficient of performance varied from 0.25 to 0.45, showing the feasibility of the mixture to be used in absorption systems operating at low generator temperatures.

Le Pierrès *et al.* (2007) evaluated a thermo-chemical discontinuous cooling prototype using the ammonia–barium chloride mixture. The results showed that temperatures as low as −30 °C can be obtained with coefficients of performance of about 0.031.

Abdelmessih *et al.* (2007) analyzed the performance of an absorption cycle operating with ethylene glycol–water as working fluids. The cycle has coefficients of performance of 0.67 and 1.2 for refrigeration and heat pump, respectively.

He and Chen (2007) reported the results of the performance of a heat-driven novel auto-cascade absorption refrigeration cycle using a mixture of R23 + R32 + R134a–DMF as its working pair. The results showed that the system can produce temperatures as low as −50 °C with generating temperatures of 163 °C.

De Lucas *et al.* (2007) investigated experimentally the mass transfer characteristics of water vapor absorption into mixtures of lithium bromide and organic salts of sodium and potassium (formate, acetate, and lactate). The authors reported that taking into account the aspect of operating range and mass transfer characteristics, the LiBr + CHO2Na (sodium formate) + water (LiBr–CHO$_2$Na = 2 by mass) may be a promising working fluid as an alternative to the LiBr + water solution.

4.4.1.3 Components

As was mentioned in the past section, in order to develop efficient absorption air conditioning systems it is important to study the heat and mass transfer of conventional and alternative mixtures in both conventional and new components. Some of the most relevant experimental studies made on components are the following.

Yoon et al. (2008) reported experimental results of the heat and mass transfer characteristics of a water–LiBr horizontal tube absorber made of small diameter tubes. The results showed that the heat and mass transfer performance of the absorber increases as the tube diameter decreases.

Sieres et al. (2008) experimentally analyzed the ammonia–water rectification in absorption systems with a 10 mm metal Pall ring packing. It was found that the height of the rectification column can be reduced up to 3 times by using the proposed Pall ring packing instead of using conventional rings.

Mohideen and Renganarayanan (2008) reported the results of experimental studies on heat and mass transfer performance of a coiled tube absorber for R134a–DMAC for an absorption cooling system. They found that the optimum overall heat transfer coefficient for R134a–DMAC solution was $726 \, W/m^2 \, K$ for a film Reynolds number of 350 and that the R134a vapor absorption rate was maximum in the normalized coil height of 0.6–1.

Zhang et al. (2006) carried out experimental research on the performance of a bubble pump with lunate channels, which provides the drive for an absorption refrigeration unit. The performance of the bubble pump is determinant in the efficiency of the absorption refrigeration system. It was shown that the performance of the bubble pump depends mainly on the driving temperature, the solution head, and the tube diameters. The elevating capability of the bubble pump with lunate channels is much better than that of conventional bubble pumps.

Bourouis et al. (2005) studied the absorption of water vapor in a wavy laminar falling film of water–(LiBr + LiI + LiNO$_3$ + LiCl) in a vertical tube at air-cooling thermal conditions. It was found that the higher solubility of the multicomponent salt solution enables the operation of the absorber at higher salt concentrations than with the conventional working fluid water–LiBr. The absorption fluxes achieved with water–(LiBr + LiI + LiNO$_3$ + LiCl) at a concentration of 64.2 wt% were around 60 % higher than those of water–LiBr at a concentration of 57.9 wt%.

Meacham and Garimella (2004) did an experimental investigation of an ammonia–water absorber that utilizes microchannel tube arrays in which liquid ammonia–water solution flows in the falling-film mode around an array of small diameter coolant tubes, while vapor flows upward through the tube array countercurrent to the falling film. The authors reported overall and solution-side heat and mass transfer coefficients and mentioned that with modifications the actual absorber, which is approximately 30 % of the surface area of the previous prototype absorber, was able to transfer duties as high as 15.1 kW, almost equalling the load of the original larger absorber.

Lee et al. (2003) analyzed both the numerical and experimental absorption process of a bubble mode absorber. It was found that at higher gas flow rates, the region of gas absorption increases. As the temperature and concentration of the input solution decrease, the region of gas absorption decreases. In addition, the absorption performance of the countercurrent flow was superior to that of concurrent. The results obtained with the model were similar to those obtained experimentally.

Islam *et al.* (2003) described the development of a novel film-inverting design concept for falling-film absorbers in which the solid surface of the absorber is segmented so that both surfaces of the falling-film are alternatively cooled in a periodic manner. A conventional tubular absorber is modified by introducing film-guiding fins between tubes to produce a film-inverting arrangement. It was shown that a maximum increase in vapor absorption rate of about 100 % can be obtained with the film-inverting design compared to the tubular absorber.

Cerezo *et al.* (2008) published an experimental study of an ammonia–water bubble absorber using a plate heat exchanger for absorption refrigeration machines. Experiments were carried out using a corrugated plate heat exchanger with three channels, where ammonia vapor was injected in bubble mode into the solution in the central channel. The authors report that the increase in pressure, solution, and cooling flow rates positively affect the absorber performance, while an increase in the concentration, cooling, and solution temperature negatively affects the absorber performance.

4.4.2   Theoretical Studies

4.4.2.1   Advanced Systems

Theoretical studies on the performance and development of advanced systems have been proposed by various authors in the last years in order to find new configurations with higher coefficients of performance, temperature lifts, or lower heat source temperatures.

Theoretical studies of hybrid cooling systems combining absorption and compression technologies have been proposed by various authors. Ludovisi *et al.* (2006) and Worek *et al.* (2003) proposed the use of a vapor recompression absorber in a double-effect absorption cooling system operating with a water–lithium bromide mixture. The authors reported that with the proposed systems higher cooling capacities and coefficients of performance 38 % higher than single-stage systems can be obtained with the proposed system. Kairouani and Nehdi (2006) studied a system operating with geothermal energy to supply a vapor absorption system cascaded with conventional compression system. The working fluids R717, R22, and R134a were selected for the conventional compression system and the ammonia–water pair for the absorption system. The results showed that the COP can be improved by 37–54 %, compared with the conventional cycle, under the same operating conditions.

Sözen and Özalp (2005) proposed another type of hybrid cooling systems utilizing an ejector in an absorption system to increase the temperature lift in order to produce refrigeration with solar energy as heat source. The absorption system operates with ammonia–water and was shown to be adequate to operate in Turkey for the most part of the year with a minimum use of the auxiliary heater.

Sabir et al. (2004) analyzed the performance of a novel heat driven refrigeration/heat pump cycle, which combines the vapor resorption and the GAX cycles. The cooling COP was found to approach unity, which is better than single-effect vapor absorption and resorption cycles, but not as good as GAX cycles but with the advantage of having a simple architecture.

Theoretical studies on the performance of multistage systems have been proposed by the following authors:

Wan et al. (2008) reported the performance of a solar air conditioning system in which a high pressure generator was added to the conventional cycle. The authors reported that the whole coefficient of performance of the proposed solar air-conditioning using mixed absorption cycle is 94.5 % higher than that of two-stage absorption cycles.

Park et al. (2008) developed models for the analysis of an ammonia GAX absorption cycle with combined cooling and hot water supply. The paper proposes new multimode GAX cycles that function in three different cooling modes. The theoretical results showed that the COP varied from 42–87 % with respect to conventional GAX cycles.

Wan et al. (2006) proposed a new two-stage solar refrigeration cycle in which the lithium bromide solution from a high pressure generator was mixed with the solution from a low pressure absorber, in order to increase more the concentration of the lithium bromide in the high pressure generator than that in the traditional two-stage refrigeration cycles. The theoretical analysis showed that with the proposed cycle the highest COP was 0.605 at heat source temperatures in the range from 75–85 °C.

Venegas et al. (2002) analyzed double and triple-stage absorption systems operating with ammonia–lithium nitrate and ammonia–sodium thiocyanate. The results showed that with the ammonia–lithium nitrate mixture for the double-stage cycle, the coefficient of performance was 0.29 at an evaporation temperature of −15 °C and a generation temperature of 90 °C. The COP for the ammonia–sodium thiocyanate mixture was a little lower.

Kaita, (2002) analyzed three kinds of triple-effect absorption cycles of parallel-flow, series-flow, and reverse-flow using a newly developed simulation program. The results showed that the parallel-flow cycle yields the highest coefficient of performance among the cycles, while the maximum pressure and temperature in the reverse-flow cycle are lower than those of other cycles.

Medrano et al. (2001) compared the potential of the organic fluid mixtures trifluoroethanol (TFE)–tetraethylenglycol dimethylether (TEGDME or E181) and methanol–TEGDME as working pairs in double-lift absorption cycles with the same cycles operating with the ammonia–water mixture. The results showed that the coefficients of performance can be up to 15 % higher with TFE–TEGDME than those obtained with ammonia–water.

Göktun and Er (2000) investigated the effects of thermal resistances and internal irreversibilities on the performance of cascaded and double-effect absorption refrigeration cycles. It was found that, under the same operation conditions, the cascaded cycle gives an increase of about 60 % in the COP and 40 % in the cooling load compared to that of a double effect cycle.

In order to increase the overall system efficiency cogeneration systems have been proposed by the following authors.

Bruno *et al.* (2008) studied the performance of biogas-driven micro gas turbine in cogeneration with absorption chillers. The study included single and double-effect commercially available chillers operating with the ammonia–water and water–lithium bromide mixtures. Micro gas turbines are fuelled with biogas and their waste heat is used to drive absorption chillers and other thermal energy users. The results showed that the best configurations are those that completely replace the existing system with a trigeneration plant that uses all the available energy.

Pilatowsky *et al.* (2007) proposed a novel cogeneration system in which the exhaust heat from a PEM fuel cell is used to operate an absorption air conditioning system operating with a monomethylamine–water mixture. The authors reported that at 80 °C the PEM fuel cell operates at its maximum efficiency and that the absorption system can operate adequately to obtain an elevated overall efficiency of the cogeneration system.

Medrano *et al.* (2006) studied the performance of a real microturbine and a single-double effect absorption chiller. The cooling system is a single-effect unit with an additional generator that recuperates the heat delivered from the microturbine. The results showed that the overall efficiency of the cogeneration system is considerable higher than that of systems operating separately.

Hwang (2004) analyzed the performance of a refrigeration system integrated with a microturbine. The waste heat from the microturbine is utilized to operate the absorption chiller. The results showed that with the proposed system the size of the microturbine can be reduced and the annual energy consumption can be reduced by up to 19 %.

4.4.2.2 Alternative Mixtures

As it was mentioned in Section 4.4.2.2 alternative mixtures to the water–lithium bromide and ammonia–water (which have been the most used for air conditioning and refrigeration, respectively) have been proposed and used by different authors. In this section some of the most relevant theoretical studies on the performance of refrigeration or air conditioning systems are presented.

Abdulateef *et al.* (2008) recently studied the performance of a solar absorption refrigeration system operating with the ammonia–lithium nitrate and ammonia–sodium thiocyanate mixtures. Their results showed that the ammonia–lithium nitrate and ammonia–sodium thiocyanate cycles give better performance than the ammonia–water cycle, not only because of higher COP values but also because rectifiers ar not required.

Chekir *et al.* (2006) analyzed the performance of butane–octane and propane–octane mixtures in an absorption chiller at temperatures lower than 150 °C and found that the coefficients of performance are about 0.63, which are almost the same than those obtained with the ammonia–water mixture.

Romero *et al.* (2005) analyzed the thermodynamic performance of a mono-methylamine–water mixture in an absorption refrigeration system and found that with this mixture better coefficients of performance can be obtained than with the ammonia–water mixture at low generation temperatures.

Rivera and Rivera (2003) published a paper on the theoretical performance of an intermittent absorption refrigeration system operating with ammonia–lithium nitrate mixture installed at the Energy Research Centre of the National University of Mexico, using a compound parabolic concentrator as generator-absorber. The results showed that with the proposed system it is possible to produce up to 11.8 kg of ice at generation temperatures around 120 °C and condensation temperatures between 40 and 44 °C. The efficiency of the system was between 0.15 and 0.4.

Romero *et al.* (2001) compared the theoretical performance of the modeling of a solar absorption system for simultaneous cooling and heating operating with water–lithium bromide and an alternative aqueous ternary mixture consisting of sodium, potassium, and cesium hydroxides in the proportions 40:36:24, respectively. The results showed that, in general, the system with the hydroxide mixture may operate with higher coefficients of performance than that with the lithium bromide mixture. It was also shown that the system with the hydroxide may operate with a higher range of temperatures.

Yoon and Kwon (1999) proposed a water–LiBr+HO(CH$_2$)$_3$OH solution and reported that the new mixture provides a crystallization limit that is 8 % higher than the conventional water–LiBr solution and coefficients of performance that are 3 % higher.

Saravanan and Maiya (1998) studied a vapor absorption refrigeration system using water as refrigerant and four binary mixtures, five ternary mixtures, and seven quaternary mixtures salts-based as absorbents. It was concluded that the water–LiCl combination is better with respect to the cut-off temperature and circulation, and the water–LiBr+LiCl+ZnCl$_2$ combination is better with respect to the coefficient of performance and efficiency ratio.

4.4.2.3 Components

In order to develop efficient sorption air conditioning systems or refrigerators, theoretical studies of heat and mass transfer of conventional and alternative mixtures are of great relevance for the optimum design of conventional components and also for the design of new more efficient components.

Although all components such as generators, absorbers, evaporators, condensers and heat exchangers are essential for the operation of sorption systems, it is a fact that the absorbers are the components are more difficult to design and their performance greatly affects the overall system performance.

Some of the most relevant theoretical studies of absorbers operating with conventional and non-conventional mixtures are the following.

Fernández-Seara et al. (2007) developed a detailed mathematical model based on mass and energy balances of an ammonia–water vertical tubular absorber cooled by air. The tubes are externally finned with continuous plate fins and the tube rows are arranged staggered in the direction of the air flow. The air is forced over the tube bank and circulates between the plain fins in cross flow with the ammonia–water mixture. The authors carried out a parametric study to investigate the influence of the design parameters and operating conditions on the absorber performance.

Chen et al. (2006) simulated and compared an innovative hybrid hollow fiber membrane absorber and heat exchanger (HFMAE) made of both porous and non-porous fibers with a plate heat exchanger falling film type absorber. The porous fibers allow both heat and mass transfers between absorption solution phase and vapor phase, while the non-porous fibers allow heat transfer between absorption solution phase and cooling fluid phase only. The results showed a substantially higher amount of absorption obtained by the proposed system by the vast mass transfer interfacial area per unit device volume provided. The application of HFMAE as the solution-cooled absorber and the water-cooled absorber in a typical ammonia–water absorption chiller allows the increase of COP by 14.8 % and the reduction of the overall system exergy loss by 26.7 %.

Venegas et al. (2005) presented a numerical procedure to design spray absorbers for ammonia–lithium nitrate absorption refrigeration systems, simulating the heat and mass transfer between solution drops and the refrigerant vapor. The results showed that about 60 % of the total mass transfer occurs during the deceleration period of the drops. This period represents 13.4 and 11.6 % of the drops' residence time inside the low and high-pressure absorbers until they reach the equilibrium state, respectively.

Tae Kang et al. (2000) analyzed a combined heat and mass transfer for an ammonia–water absorption process and carried out the parametric analysis to evaluate the effects of important variables such as heat and mass transfer areas on the absorption rate for two different absorption modes: falling film and bubble modes in a plate heat exchanger with an offset strip fin in the coolant side. It was found that the local absorption rate of the bubble mode was always higher than that of the falling film model, leading to an about 48.7 % smaller size.

References

Abdelmessih AN, Abbas M, Al-Hashem A, Munson J (2007) Ethylene glycol/water as working fluids for an experimental absorption cycle. Experimental Heat Transfer 20(2):87–102

Abdulateef JM, Sopian K, Alghoul MA (2008) Optimum design for solar absorption refrigeration systems and comparison of the performances using ammonia-water, ammonia-lithium nitrate and ammonia-sodium thiocyanate solutions. Int J Mech Mater Engg 3(1):17–24

Ajib HS, Gunther W (2007) Investigation results of an absorption refrigeration machine operated solar thermally for cooling and air conditioning under using a new work solution. In: 2nd Int Conference on Solar Air-Conditioning, Tarragona, pp. 510–516

Arivazhagan R, Saravanan S, Renganarayanan S (2006) Experimental studies on HFC based two-stage half effect vapour absorption cooling system. Appl Therm Eng 26(14–15): 1455–1462

Bourouis M, Vallès M, Medrano M, Coronas A (2005) Absorption of water vapour in the falling film of water – (LiBr + LiI + LiNO3 + LiCl) in a vertical tube at air-cooling thermalconditions. Int J Therm Sci 44(5):491–498

Bruno JC, Ortega-López V, Coronas A (2008) Integration of absorption cooling systems into micro gas turbine trigeneration systems using biogas: Case study of a sewage treatment plant DOI: 10.1016/j.apenergy 2008.08.007

Caeiro JA (2008) Experimental testing of an innovative lithium-bromide water absorption refrigeration cycle coupled with ice storage. Int J Low Carbon Technol 3(1):59–69

Cerezo J, Bourouis M, Vallès M, Coronas A, Best R (2008) Experimental study of an ammonia-water bubble absorber using a plate heat exchanger for absorption refrigeration machines. Applied Thermal Engineering DOI: 10.1016/ j.applthermaleng 2008.05.012

Chekir N, Mejbri K, Bellagi A (2006) Simulation of an absorption chiller operating with alkane mixtures. Int J Refrig 29(3):469–475

Chen J, Chang H, Chen SR (2006) Simulation study of a hybrid absorber-heat exchanger using hollow fiber membrane module for the ammonia-water absorption cycle. Int J Refrig 29(6):1043–1052

De Lucas A, Donate M, Rodríguez JF (2007) Absorption of water vapor into new working fluids for absorption refrigeration systems. Ind Eng Chem Res 46(1):345–350

Ezzine NB, Mejbri KH, Bellagi A (2007) Solar driven hydrocarbon operated absorption-difussion machine. In: 2nd Int. Conference on Solar Air-Conditioning, Tarragona, pp. 497–502

Fernández-Seara J, Uhía FJ, Sieres J (2007) Analysis of an air cooled ammonia-water vertical tubular absorber. Int J Therm Sci 46(1):93–103

Glebov D, Setterwall F (2002) Experimental study of heat transfer additive influence on the absorption chiller performance. Int J Refrig 25(5):538–545

Göktun S, Er ID (2000) Optimum performance of irreversible cascaded and double effect absorption refrigerators. Appl Energ 67(3):265–279

Gómez VH, Vidal A, Best R, García-Valladares O, Velázquez N (2008) Theoretical and experimental evaluation of an indirect-fired GAX cycle cooling system. Appl Therm Eng 28 (8–9):975–987

Häberle A, Luginsland F, Zahler C, Berger M, Rommel M, Henning HM, Guerra M, De Paoli F, Motta M, Aprile M (2007) Alinear concentrating Fresnel collector driving a NH₃-H₂O absorption chiller. In: 2nd Int. Conference on Solar Air-Conditioning, Tarragona, pp. 662–667

He Y, Chen G (2007) Experimental study on a new type absorption refrigeration system. Taiyangneng Xuebao/Acta Energiae Solaris Sinica 28(2):137–140

Hwang Y (2004) Potential energy benefits of integrated refrigeration system with microturbine and absorption chiller. Int J Refrig 27(8):816–829

Islam MR, Wijeysundera NE, Ho JC (2003) Performance study of a falling-film absorber with a film-inverting configuration. Int J Refrig 26:909–917

Jakob U, Pink W (2007) Development and investigation of an ammonia/water absorption chiller – chillii® PSC – for solar cooling system. In 2nd Int Conference on Solar Air-Conditioning, Tarragona, pp. 440–445

Kaita Y (2002) Simulation results of triple-effect absorption cycles. Int J Refrig 25(7):999–1007

Le Pierrès N, Mazet N, Stitou D (2007) Experimental results of a solar powered cooling system at low temperature. Int J Refrig 30(6):1050–1058

Lee JC, Lee KB, Chun BH, Lee CH, Ha JJ, Kim SH (2003) A study on numerical simulations and experiments for mass transfer in bubble mode absorber of ammonia and water. Int J Refrig 26(5):551–558

Llamas SU, Herrera JV, Cuevas R, Gómez VH, García-Valladares O, Cerezo J, Best R (2007) Development of a small capacity ammonia-lithium nitrate absorption refrigeration system. In: 2nd Int Conference on Solar Air-Conditioning, Tarragona, pp. 470–475

Ludovisi D, Worek WM, Meckler M (2006) Simulation of a double-effect absorber cooling system operating at elevated vapor recompression levels. HVAC&R Res 12(3):533–547

Meacham JM, Garimella S (2004) Ammonia-water absorption heat and mass transfer in microchannel absorbers with visual confirmation. ASHRAE Trans 110(1):525–532

Medrano M, Bourouis M, Coronas A (2001) Double-lift absorption refrigeration cycles driven by low-temperature heat sources using organic fluid mixtures as working pairs. Appl Energ 68(2):173–185

Medrano M, Mauzey J, McDonell V, Samuelsen S, Boer D (2006) Theoretical analysis of a novel integrated energy system formed by a microturbine and an exhaust fired single-double effect absorption chiller. Int J Thermodyn 9(1):29–36

Meyer J (2008) What solar cooling costs. Sun and Wind Energy 1:82–84

Mohideen ST, Renganarayanan S (2008) Experimental studies on heat and mass transfer performance of a coiled tube absorber for R134a-DMAC based absorption cooling system. Heat Mass Transfer 45(1):47–54

Moser H, Rieberer R (2007) Small-capacity ammonia/water absorption heat pump for heating and cooling used for solar cooling. In: 2nd Int Conference on Solar Air conditioning. Tarragona, pp 56–61

Mugnier D (2008) phttp://lmora.free.fr/task38/pdf/matin/Mugnier.pdf. Accessed 8 Oct 2008

Mugnier D, Hamadi M, Le Denn A (2008) Water chillers-closed systems for chilled water production (small and large capacities). Int Seminar on Solar Air-Conditioning, Munich, pp. 31–37

Muthu V, Saravanan R, Renganarayanan S (2008) Experimental studies on R134a-DMAC hot water based vapour absorption refrigeration systems. Int J Therm Sci 47(2):175–181

Park CW, Koo J, Kang YT (2008) Performance analysis of ammonia absorption GAX cycle for combined cooling and hot water supply modes. Int J Refrig 31(4):727–733

Pilatowsky I, Romero RJ, Isaza CA, Gamboa SA, Rivera W, Sebastian PJ, Moreira J (2007) Simulation of an air conditioning absorption refrigeration system in a co-generation process combining a proton exchange membrane fuel cell. Int J Hydrogen Energ 32(15):3174–3182

Rivera CO, Rivera W (2003) Modeling of an intermittent solar absorption refrigeration system operating with ammonia–lithium nitrate mixture. Sol Energ Mat Sol C 76(3):417–427

Romero RJ, Guillen L, Pilatowsky I (2005) Monomethylamine-water vapour absorption refrigeration system. Appl Therm Eng 25(5-6):867–876

Romero RJ, Rivera W, Pilatowsky I, and Best R (2001) Comparison of the modeling of a solar absorption system for simultaneous cooling and heating operating with an aqueous ternary hydroxide and with water/lithium bromide. Sol Energ Mat Sol C 70(3):301–308

Sabir HM, Chretienneau R, El Hag YBM (2004) Analytical study of a novel GAX-R heat driven refrigeration cycle. Appl Therm Eng 24(14–15):2083–2099

Saravanan R, Maiya MP (1988) Thermodynamic comparison of water-based working fluid combinations for a vapour absorption refrigeration system. Appl Therm Eng 18(7):553–568

Sieres J, Fernández-Seara J, Uhía FJ (2008) Experimental analysis of ammonia-water rectification in absorption systems with the 10 mm metal Pall ring packing. Int J Refrig 31(2):270–278

Sözen A, Özalp M (2005) Solar-driven ejector-absorption cooling system. Appl Energy 80(1):97–113

Sun ZG, Guo KH (2006) Cooling performance and energy saving of a compression-absorption refrigeration system driven by a gas engine. Int J Energ Res 30(13):1109–1116

Tae Kang Y, Akisawa A, Kashiwagi T (2000) Analytical investigation of two different absorption modes: falling film, and bubble types. Int J Refrig 23(6):430–443

Troi A, Napolitano A, Sparber W (2008) Overview of solar cooling systems for commercial buildings. Int. Seminar on Solar Air-Conditioning, Munich, pp. 81–91

Venegas M, Izquierdo M, De Vega M, Lecuona A (2002) Thermodynamic study of multistage absorption cycles using low-temperature heat. Int J Energ Res 26(8):775–791

Venegas M, Izquierdo M, Rodríguez P, Nogueira JI (2005) Design of spray absorbers for LiNO$_3$-NH$_3$ absorption refrigeration systems. Atomization Spray 15(4):439–456

Wagner TC, Rog L, Jung SH (2008) Development of a simultaneous cooling and heating absorption chiller for combined heat and power systems. ASME Int Mechanical Engineering Congress and Exposition. Proceedings Vol 15, pp. 73–78

Wan Z, Shu S, Hu X (2006) Novel high-efficient solar absorption refrigeration cycles. J of Huazhong University of Science and Technology 34(9):85–87

Wan Z, Shu S, Hu X, Wang B (2008) Research on performance of mixed absorption refrigeration for solar air-conditioning. Front Energ Power Eng China 2(2):222–226

Worek WM, Ludovisi D, Meckler M (2003) Enhancement of a double-effect absorption cooling system using a vapor recompression absorber. Energy 28(12):1151–1163

Yaxiu G, Yuyuan W, Xin K (2008) Experimental research on a new solar pump-free lithium bromide absorption refrigeration system with a second generator. Sol Energy 82(1):33–42

Yoon JI, Kwon OK (1999) Cycle analysis of air-cooled absorption chiller using a new working solution. Energy 24(9):795–809

Yoon JI, Phan TT, Moon CG, Lee HS, Jeong SK (2008) Heat and mass transfer characteristics of a horizontal tube falling film absorber with small diameter tubes. Heat and Mass Transfer 44(4):437–444

Zetzsche M, Koller T, Brendel T, Muller-Steinhagen H (2007) Solar cooling with an ammonia/water absorption chiller. In: 2nd Int Conference on Solar Air-Conditioning, Tarragona, pp. 536–541

Zhang L, Wu Y, Zheng, H, Guo J, Chen D (2006) An experimental investigation on performance of bubble pump with lunate channel for absorption refrigeration system. Int J Refrig 29(5):815–822

Chapter 5
Sorption Refrigeration Systems

5.1 Introduction

The knowledge of the basic principles of thermodynamics allows us to understand the conditions and necessary limitations in order to transform heat in work, transferring heat from a thermal source of high temperature to a smaller one. Thermal machines work under this principle, however, there are machines that consume work (external) and produce heat, that is to say in the inverse sense of a thermal machine operation according to the cycle of Carnot, this it is the case of a refrigerating machine. A particular case is refrigeration cycles based on the sorption process, which operate with thermal energy and consume their own work that they self-produce, this being the coupling among a thermal machine and a refrigeration machine. There are great variety of sorption refrigeration systems, in general those of absorption and adsorption cycles.

In this chapter, the introduction of the sorption theory, its applications to refrigeration thermodynamic cycles, and its efficiencies are presented and analyzed, as well as the different possibilities of work fluids for diverse applications, showing different examples of refrigeration cycles.

5.2 Thermodynamic Principles

5.2.1 Heat to Work Energy Conversion

Among the very varied forms of energy conversion, one is to transform the thermal potential into usable work. The simplest form of achieving this is by means of a thermal machine.

It is necessary to distinguish reversible transformations from irreversible ones. A reversible transformation can be considered as a continuous series of equilibrium states. When changing infinitely little the equilibrium factors, the process can be carried out in one sense or in the opposite; expansion and compression and vaporization and condensation, for example. There are irreversible transformations in nature that are always developed in a single sense. To understand this transformation from the theoretical point of view, the application of the principles of thermodynamic is indispensible.

It is important to consider the two enunciations by of Clausius, (1850) and Lord Kelvin, (1851). Clausius says "The step of heat from a cold body to one hot, is never carried out in a spontaneous way or it is never carried out, without compensation". The terms spontaneity and compensation mean that the transformation is not possible, only if this is bound one to a modification that is carried out in the external medium at the same time.

When a system is in contact with a single thermal source at constant temperature, this transformation is called mono-thermal. According to Kelvin's enunciated, it is impossible to obtain work. In order to transform the thermal energy into work, it is necessary to have a difference of temperature, therefore to have two thermal sources.

Sadi Carnot (1824), was the first to establish a precise relationship between heat and work, giving some necessary conditions: (1) mechanical work cannot exist if it is not transferred from a body of a bigger temperature to one of a smaller temperature and the work will be bigger, as this difference is bigger, (2) the maximum work is given when the changes are reversible, including the concept of "reversible cycle", which consists of two isothermal and two adiabatic processes.

5.2.1.1 Carnot's Di-thermal Cycle, the Thermal Machine

When a system during a cycle is successively in thermal contact with two sources of heat, a hot source S_{T1}, at temperature T_1, and a cold source S_{T2} at temperature T_2, with $T_1 > T_2$, it is known as a di-thermal cycle.

A certain cycle contains two mono-thermal transformations during which the system exchanges heat with the corresponding sources. Considering the hypothesis that the cycle should not contain other sources of temperature, the transfer of the thermal contact of one of the sources to the other one needs to be isolated thermally from the system; this means that two operations exist intermediately adiabatically: an adiabatic operation that carries out the step of the thermal contact from S_{T1} to S_{T2} and another adiabatic operation that carries out the passage of the thermal contact from S_{T2} to S_{T1}.

The di-thermal cycle, contains four successive and different operations: (1) a mono-thermal operation at temperature T_1, during which the system exchanges a certain quantity of heat Q_1, with the source of heat S_{T1}; (2) an adiabatic operation that interrupts the thermal contact with S_{T1} and takes the system to the thermal contact with the source S_{T2}; (3) a mono-thermal operation at temperature T_2, during which the system exchanges a certain quantity of heat Q_2, with the source S_{T2}; and (4) finally an adiabatic operation that interrupts the thermal contact with S_{T2} and that puts in thermal contact with S_{T1} under such conditions that the initial state of the system is recovered. When the four operations that constitute the di-thermal cycle are reversible, the cycle takes the name of Carnot's cycle.

According to the corollary of the theorem of Carnot, the performance of irreversible di-thermal machine is inferior to that of a reversible machine using the same sources of heat. The sense of the cycle for the reversible machine can be reinvested. If two machines are coupled, an irreversible one working in direct sense and other reversible one in inverse sense, the exchange of heat is annulled with one of the sources. The two coupled machines realize an irreversible mono-thermal cycle, exchanging heat with a single source. This can demonstrate that the performance of an irreversible machine is inferior to that of a reversible machine, having the same sources S_{T1} and S_{T2} at the same temperatures T_1 and T_2.

The concept of reversible thermal machine corresponds then to an ideal system, to a perfect machine that enables achievement of the best benefit of the two given sources of heat for the work production. The efficiency of the reversible machine corresponds to a maximum limit value of the efficiency of the real machines, and this limit efficiency does not depend absolutely on the nature of the machine or on the nature of the work fluid. This efficiency has a universal meaning and it depends only on the temperatures (Cardwell 1971).

The Carnot cycle is an idealized energy conversion cycle. Figure 5.1 shows a Carnot cycle for work generation on a temperature-entropy diagram. The process line 1–2 represents the isothermal addition of heat Q_2 at temperature T_2. Line 2–3 represents the isentropic production of work, line 3–4 represents the isothermal dissipate heat Q_1 at temperature T_1, and finally the line 4–1 the isentropic input of work. If all processes are assumed reversible, then the area enclosed by 1–2–3–4 represents the net amount of work produced W, and the area 3–4–5–6 represents the amount of thermal energy Q_1 dissipated by the cycle, when 5 and 6 are at $T = 0\,\text{K}$ (zero absolute). The addition of the two areas, 1–2–6–5, is the amount of Q_2 supplied to the cycle, according to the first law of thermodynamics. The process of work production follows a clockwise direction on T-s diagram.

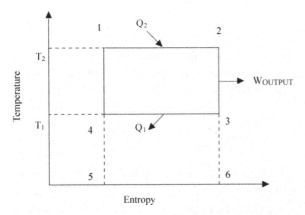

Figure 5.1 The Carnot cycle for work production on a temperature-entropy diagram

According to the first law of thermodynamics

$$Q_2 = Q_1 + W \tag{5.1}$$

The efficiency η for work generation is defined as the amount of work produced divided by the amount of heat supplied at the high temperature, then

$$\eta = \frac{W}{Q_2} = \frac{Q_2 - Q_1}{Q_2} \tag{5.2}$$

The second law of thermodynamics states for the Carnot cycle (Figure 5.2), that for reversible process the net entropy production is zero, so that

$$s_2 - s_1 = \frac{Q_2}{T_2} - \frac{Q_1}{T_1} = 0 \tag{5.3}$$

Equation 5.3 can be expressed in terms of absolute temperatures

$$\eta = \frac{T_2 - T_1}{T_2} \tag{5.4}$$

This expression is termed as Carnot efficiency factor (η) for work production.

If in the Carnot cycle it is considered that the direction of all processes is reversed as compared to the work generation, this is operated as a heat pump cycle. Figure 5.2 shows the Carnot cycle of heat pump.

The Carnot cycle for heat pump can be also represented on the T-s diagram, (IIR 1958) where the temperatures are chosen such that heat Q_0 is added at T_0 along the process 3–4. The working fluid is compressed isotropically 4–1 and heat Q_1 is dissipated at T_1, 1–2 process, and the fluid expanded isotropically, 2–4 process. The amount of work input required for the cycle is represented by the area 4–3–2–1 and the amount of heat absorbed by the area 3–4–6–5. The sum of both areas, 1–2–5–6 represents the amount of heat dissipate at T_1.

The performance of a heat pump is defined by the ratio of the heat available at high temperature, divided by the net work requirement, where COP is applied

$$COP = \frac{T_1}{T_1 - T_0} \tag{5.5}$$

The Carnot cycle of Figure 5.2, can be used for cooling or refrigeration applications. For refrigeration, the heat removed at T_0 is of interest and the COP is defined as the ratio of the cooling capacity Q_0 over the work input, in this case the tem COP_R is used.

$$COP_R = \frac{T_0}{T_1 - T_0} \tag{5.6}$$

For the coefficients of performance of the heat pump and refrigeration cycle, the following relationship can be used:

$$COP_R + 1 = COP \tag{5.7}$$

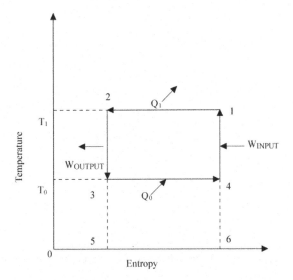

Figure 5.2 The Carnot cycle for a heat pump (refrigeration cycle) on a temperature–entropy diagram

In order to describe a sorption heat pump with Carnot cycles, both the work production and refrigeration cycles may be combined into one thermal machine. It is assumed that the amount of work produced by the first cycle (Figures 5.1 and 5.2) is identical to the amount of work required by the second cycle.

The Carnot closed cycle is a reference theoretical cycle, which can be applied to refrigeration cycles, including both mechanical compression and sorption. It is a basic comparison for any type of refrigerating machine. The comparison and the determination of the efficiencies can be expressed equally in the case of the sorption to the interior or the external regime.

The interior regime of a refrigeration cycle refers exclusively to the changes of state of the refrigerant that goes through the closed cycle. The external regime refers to the transformations of the ambient fluid (water or air), which is the part of the system where the heat is released.

Figure 5.3 represents the Carnot cycle of a refrigeration sorption machine on a temperature-entropy, T-s, diagram. From a thermodynamic point of view a sorption refrigerating machine is considered as a thermal machine coupled to a compressor for the production of cold.

If the two machines work following the Carnot cycle reversibly, the area of the rectangle 1–2–3–4 represents the work given by the thermal machine, and the area of the rectangle 5–6–7–8 represents the work absorbed by the refrigeration machine (compressor).

In this kind of disposition, no external mechanical work appears (perfect case), since the entire work given by the thermal machine is totally used for the operation of the compressor. In consequence, the areas of the rectangles 1–2–3–4 and 5–6–7–8 should be the same.

In the T-s diagram the heat quantities involved are represented by means of surfaces. To the exterior only the quantities of heat appear, as being able to represent

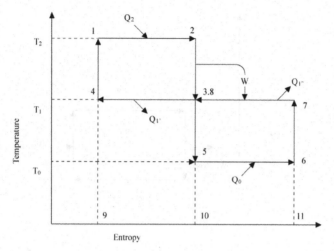

Figure 5.3 Temperature-entropy diagram of the work production of a sorption refrigerating machine

in the diagram *T-s*, by means of surfaces. The area of the rectangle 1–2–10–9 represents the heating supplied at the high temperature T_2, and the area of the rectangle 5–6–7–8 represents the work absorbed by the compressor to an interme-diate temperature T_1, while the rectangle 6–5–10–11 represents the refrigerating power to the temperature T_0.

5.2.2   Vapor Compression Refrigeration Cycle

The compression refrigeration cycle is an example of an inverse Carnot cycle, where in order to transfer heat from a source of low temperature to one of high temperature, external work is required. In this case, it is provided by the compres-sor and the four described transformations are carried out.

Figure 5.4 represents compression of refrigerating machine to single stage. This is formed by (1) an evaporator (*E*) where the refrigerant is evaporated under evaporation pressure (P_E). The vapors formed are, in general, superheated to the exit of this exchanger; (2) a mechanical compressor (*C*) that is suctioned under the pressure, (P_E). The superheated vapors are then are compressed to the pressure (P_C) corresponding to the temperature of saturation, usually referred to as the condensation temperature; (3) a condenser (*C*) where the superheated vapor is liquefied to T_C and the liquid one can be sub-cooled. The condenser can be cooled with environmental fluids such as water and air; and (4) an expansion valve E_V, which receives the liquid refrigerant and is expanded from P_C to the pressure P_E, being an isenthalpic expansion where the quality of the vapor diminishes. The liquid passes again to the evaporator and this way a new refrigeration cycle begins.

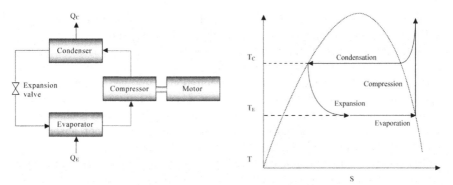

Figure 5.4 Vapor compression refrigeration cycle

5.3 Sorption Processes

5.3.1 Introduction

Sorption is a physico-chemical phenomenon that consists in fixing an element or compound on a liquid or solid phase, exercising a suction effect with a physical or chemical character, depending on the nature of the phases involved phases.

In the sorption process the interaction mechanisms are varied and complex, in most cases they are characterized by a mass diffusion of the element to transfer, from the interior of its own phase toward the surface of the other one. If a superficial interaction exists among these two phases, for example, by means of attractive forces of physical character, it is known as an adsorption process. In this case, apart from physical adsorption, there is chemical adsorption or chemical-sorption. In the adsorption, there are no physical deformations of the phases and as the active surface of the adsorbent is bigger, the fixed quantity of the phase absorbate will be more important.

There is another sorption process, which consists on the solubilization of the working fluid in the sorbent, well-known as absorption. In this case, the fluid is spread to the interior of the phase of the sorbent, being integrated to the phase of the absorbate. There are cases of absorption, where once the fluid is dissolved this begins to react chemically, forming new compounds; this is the case of the thermal-chemical refrigeration system.

In most sorption processes, the contact among the phases dissipates a specific amount of heat, as in the cases of adsorption and absorption with or without chemical reaction. Several interaction possibilities exist among the phases. In the case of the adsorption, for example, gas or vapor can be fixed, both in the surface of a liquid and a solid, although the case of solids is more frequent. Some examples of this are: the adsorption of water vapor in different types of adsorbents, such as activated coal, silica gel, zeolitic structures, *etc.*

Both in absorption and in adsorption, the adsorbents can interact with the liquid and solid phases and in both cases a chemical interaction among the phases is possible. In this case, a fundamental difference exists in the character of the isotherms. In the case of equilibrium between a liquid and a vapor, they have a continuous function for a wide interval of concentrations. However, in the case of the equilibrium solid-vapor the isotherms present a series of steps towards constant pressure when a concentration change is present. This phenomenon is characteristic of thermochemical absorption (solid absorption of solid-vapor absorption), where different compounds are formed at different concentrations.

The sorption refrigeration cycle (SRC) presents many similarities with the vapor compression cycle since both present similar operations, for example, condensation, expansion, and evaporation of a pure component.

5.3.2   The Sorption Refrigeration Cycle

The main difference between the cycles consists in the process of compression of the vapor saturated to superheated conditions. The cycle of mechanical compression uses a compressor as long as the cycles for sorption, like the one mentioned, it takes advantage of certain physical-chemical phenomena characteristics of the interaction between the two phases. This phenomenon includes the surface interactions, the diffusion and transfer of mass of a phase into another, and the possible chemical reactions among the compounds that form the phases (Figure 5.5).

A cycle for sorption consists of the three processes common to the fluid. In the refrigeration cycle for compression they are: condensation (condenser), expansion (expansion valve), and evaporation (evaporator).

In the sorption cycle, the compressor is substituted by two processes characteristic of the sorption phenomena: the sorption that is carried out in the sorber (adsorber or absorber, depending on the nature and the sorption process) and the phenomenon of desorption that consists of the refrigerant extraction, which is carried out in the desorber or generator.

The desorption process, is of endothermic nature, contrary to the sorption process (absorption or adsorption), which is exothermic. A specific quantity of thermal energy is required to liberate the refrigerant, almost as much as that required for recombination.

Based on the above it is possible to summarize that in an SRC, the work fluid is received (refrigerant) as saturated vapor at low pressure and it is put in contact with another phase or compound (liquid or solid). When the refrigerant comes into contact with any of these phases, a sorption phenomenon (absorption, adsorption or chemical-sorption) takes place and this is added to a solid or a liquid, dissipating an amount of heat of sorption Q_{SO} at T_{SO} and P_{SO}. In the desorber the superheated vapor is liberated from the solid or liquid sorbent when an amount of heat is added Q_{DS} at T_{DS} and P_{DS}. If P_{SO} is the low pressure level and P_{DS} is the high pressure level, a single-stage SRC cycle is obtained.

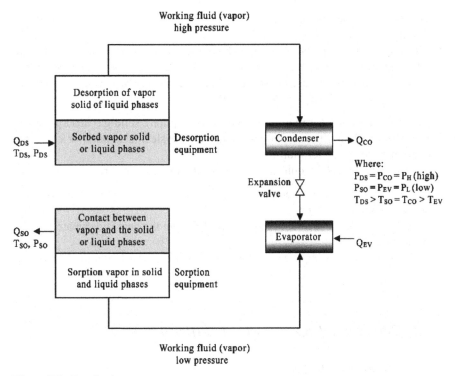

Figure 5.5 Sorption heat pump

In the case of the sorption systems the following aspects should be considered:

1. The phenomena involved in the sorption, as well as the conditions of pressure, temperature, concentration, or saturation to which they are carried out, are considered in a state of thermodynamic equilibrium among the phases.

2. In its simpler configuration, the SRC operates in an intermittent process for solid-gas systems. This is due to the mechanical difficulty associated to the transport of the solid from one component to another. In intermittent cycles the sorber and the desorber are the same and they exchange their functions in a periodic process (for example, a recipient operates as desorber, liberating the work fluid and as sorber, fixing the vapor). Once the refrigerant has been desorbed and has circulated through the condenser, the expansion valve, and the evaporator, it returns to the same recipient. It operates like sorber only in this case and the work fluid, starting from a sorption phenomenon, is added to the present phase. For liquid-gas systems, the cycle can operate as a continuous process, a pump and a valve being integrated according to the solution of the hydraulic loop between the sorber and desorber or vapor.

3. SRC operates thermally. This means that the work required by the cycle to operate is by means of thermal energy. The work of the pump required for the

circulation of the solution in an absorption cycle consumes a small electric power.

4. SRS requires thermal energy at high level temperature in the desorber and at an intermediate temperature level in which one type of heat is dissipated in the condenser. This is because the sorption cycle operates at two different levels of pressure and three levels of temperature.

Classical SRCs are: sorption liquid-gas (or absorption), solid-gas absorption (or thermo-chemical), and the sorption solid-gas (or physical adsorption or chemical adsorption or chemical-sorption).

For its operation SRC requires at least three thermal sources (tri-thermal cycle):

1. A thermal source at low temperature T_E where the heat is extracted from the space or body for cooling.
2. A thermal source at medium temperature T_C where the heat is dissipated to the environment.
3. A thermal source at high temperature T_G, where the heat is supplied to assure its operation.

5.3.3 Sorption Refrigeration Cycle Efficiency

5.3.3.1 Theoretical Efficiency

From the thermodynamic point of view it is convenient to consider a tri-thermal refrigeration system as a machine constituted by a thermal motor (MT) working with a temperatures range between T_G and T_C, and a refrigerating machine (MF) operating between T_E and T_C (Niebergall 1959).

Considering the two principles of thermodynamics for the Carnot cycle applied to sorption systems and from the conservation principle:

$$Q_D + Q_E = Q + Q_S \tag{5.8}$$

Where Q_D, Q_E, Q_C, and Q_S represent the heat supply to the desorber (vapor generation) at the highest temperature, the heat absorbed at the lower temperature (evaporator), the dissipated heat, during the liquefaction of the vapor (condenser), and the heat dissipated due to the sorption process, respectively. The condensation and sorption heats are dissipated to the intermediate temperature T, supposing that T_C and T_S are the same.

From the second principle, the entropic balance is

$$S_D + S_E = S_0 + S_S \tag{5.9}$$

With

$$\frac{Q_D}{T_D} + \frac{Q_E}{T_E} = \frac{Q}{T} + \frac{Q_S}{T_S} \tag{5.10}$$

Solving for Q_E

$$\frac{Q_E}{Q_D} = \frac{1}{\frac{1}{T_E} - \frac{1}{T}}\left[\left(\frac{1}{T} - \frac{1}{T_G}\right) - \frac{Q_S}{Q_D}\left(\frac{1}{T_E} - \frac{1}{T_S}\right)\right] \tag{5.11}$$

If the hypothesis that the sorption and desorption heat is constant and $Q_S = Q_D$ is considered, then the following holds.

For an interior regime, the theoretical thermal effect, (η_{In}) is

$$\eta_{In} = \left(\frac{Q_E}{Q_D}\right) = \frac{\left(\frac{1}{T_S} - \frac{1}{T_D}\right)}{\left(\frac{1}{T_E} - \frac{1}{T}\right)} \tag{5.12}$$

In the case of the external regime, η_{Ex}

$$\eta_{Ex} = \frac{\left(\frac{1}{T_{RF}} - \frac{1}{T_{CF}}\right)}{\left(\frac{1}{T_{SF}} - \frac{1}{T_{HF}}\right)} \tag{5.13}$$

where, T_{RF}, T_{CF}, T_{SF}, and T_{HF} are the lowest temperature in the refrigerant fluid or in the body to be cooled, the temperatures of cooling media (water or air) to the exit of the condenser and the sorber, and the temperature of the heating fluid, respectively.

The thermal efficiency of the Carnot cycle is referred to in absolute temperatures and it does not consider the properties of the work fluid or the irreversibilities due to the processes of transfer and mass transfer. If two machines work under the same conditions of temperature, they will have the same efficiency.

5.3.3.2 The Coefficient of Performance

By definition the coefficient of performance is represented as

$$COP = \frac{|Q_0|}{|Q_G|} = \frac{\text{cooling effect}}{\text{energy supplied}} \tag{5.14}$$

If the work W supplied by the thermal motor (TM), which is absorbed by the refrigerating machine (RM), is included, then

$$COP = \frac{|Q_0|}{|W|} \cdot \frac{|W|}{|Q_G|} \tag{5.15}$$

The first term represents the coefficient of efficiency of the di-thermal refrigeration cycle receiving mechanical energy and the second the performance of the thermal motor producing work. Therefore, the efficiency of a tri-thermal refrigeration system is represented by the following equation:

$$COP_{TRS} = COP_{RM}COP_{TM}$$

(5.16)

If the two coupled machines are considered perfect (ideals), then

$$COP_{TES, ideal} = COP_{RM.ideall}COP_{TM, ideal}$$

(5.17)

In terms of thermodynamic temperatures

$$COP_{TRS, ideal} = \frac{T_E}{T_C - T_E} \cdot \frac{T_G - T_C}{T_G}$$

(5.18)

5.3.4 Sorption Work Fluids

The work fluids in an SRC are the refrigerant and their liquid or solid sorbents, and their selection if very important.

5.3.4.1 Refrigerants

The refrigerant is a substance that is able to produce a cooling effect on a space or a body, which flows internally into a refrigeration cycle. In the case of cold production by vaporization, these substances should have a boiling temperature at normal pressure, lower than room temperature.

In general, the refrigerant is fluids that work in close cycles in the cooling machines. They be inorganic, for example, water, ammonia, amines, *etc.*, or organic, for example, hydrocarbons such as methane, propane, ethane, ethylene, *etc.*, and halogen hydrocarbons, particular the fluorocarbons R12, R22 R13B1, and R501 (azeotropic mixture). An important aspect in the case of SRCs is that the refrigerant fluids should be very environmentally friendly with the minimum atmospheric impact.

Selection of the Refrigerant

For each of the various methods of cold production for certain operation conditions there are one or several appropriate refrigerants that guarantee good efficiency and security. In connection with their chemical and physical properties there are certain minimum conditions and properties that should be satisfied, such as:

1. The refrigerant should not combine, be mixed, or react with the materials used for the construction of the refrigerating machine.
2. The refrigerant should be chemically stable.

3. The refrigerant should not have any kind of chemical transformation, in the domain of temperatures and operation pressures.
4. The refrigerant should not be toxic.
5. The refrigerant should not be explosive or inflammable.
6. It should be easy to detect refrigerant leaks.
7. It should not react with the lubricant.
8. The evaporation pressure should be larger than the atmospheric pressure.
9. The condensation pressure should be low.
10. There should be a high specific refrigerating power.
11. The refrigerants should be of low cost and easily available.

5.3.4.2 Sorbents

The selection of sorbents is based on certain properties, referring to their use in SRC, such as:

1. The sorbent should have a high affinity with the refrigerant, in order to reduce the amount of refrigerant circulating in the cycle, However, if this affinity is very large, it will be necessary to supply an important amount of energy for the desorption of the refrigerant.
2. The vapor pressure at the required generation temperature must be negligible or lower than the vapor pressure of the refrigerant.
3. The sorbent should remain in its liquid state during the entire cycle, in order to avoid crystallization. The chemical stability should be good, non-corrosive, and non-toxic.
4. *The specific heat should be small to avoid thermal losses with a higher thermal conductivity and lower viscosity and superficial tension in order to improve the heat transfer and the sorption process.*
5. The sorbent should be less volatile than the refrigerant, in order to facilitate its separation in the generator. If this is not possible, it will be necessary to integrate a rectifier for the sorbent vapor separation.

5.3.4.3 Refrigerant-sorbent Systems

For the optimal operation of refrigeration cycles based on the sorption principle, many studies have been developed in order to establish the good characteristics of the refrigerant-absorbent system (Hainsworth 1944, Buffington 1949, Eding and Brady 1961, Andrews 1964, Rush *et al.* 1967, Macriss 1979, Macriss *et al.* 1988, Mansoori and Patel 1979, Tyagi and Rao 1984).

The performance of the SRC depends on the choice of the appropriate pair of refrigerant and absorbent. In general, the selection of fluids is based on the following considerations: (1) chemical and physical properties of the fluids and (2) the acceptability range for certain thermophysical and thermodynamic properties of the fluids.

The main characteristics of a refrigerant-sorbent pair for an absorption refrigeration system (ARS), are: (1) low viscosity of the solution under operating conditions, in order to reduce the pumping work, (2) the freezing points of the liquids should be lower than the lowest temperature in the specific cycle; they should have good chemical thermal stability; and (3) the components should be non-corrosive, non-toxic and non-flammable properties.

In order to analyze the potential of the refrigerant-sorbent pair it is important to know the properties of the pure refrigerant and sorbent; the thermodynamic properties of their solutions are critical in determining the suitability of new fluids for ARS. The pair selected should satisfy two main thermodynamics requirements: (1) high solubility of the refrigerant in the sorbent and (2) a larger difference in boiling points of the sorbent and the refrigerant.

Based on the above, an important number of refrigerant-sorbents have been proposed, for example by Andrews (1964), Rush *et al.* (1967), Hensel and Harlowe (1972), and Raldow (1982).

5.4 Absorption Refrigeration Systems

5.4.1 Introduction

The absorption refrigeration cycle is a particular case of sorption systems, where two phases participate, in general, liquid-vapor and solid-vapor (thermal-chemical refrigeration).

Of all the available thermodynamic cycles for the production of cold, the tri-thermal sorption system is used in the application of energy at low enthalpy applications, such as *solar or heat industrial waste*, and in particular, the liquid-gas and solid-gas absorption and solid-gas adsorption in continuous and intermittent operations.

5.4.2 Working Substances

An important number of refrigerant-sorbent have been proposed (Andrews 1964, Rush *et al.* 1967, 1972, Raldow 1982).

The sulfuric acid–water solutions used in the early beginnings turned out to be corrosive and poisonous. However, in spite of the many advantages of the ammonia–water mixture, it also has some disadvantages. It cannot be avoided that a small amount of water gets into the evaporator, which absorbs ammonia and prevents it from evaporation and the water is solidified. The other aspect concerns the higher operating pressures. Furthermore the maximum temperature generation is limited, where, above 180 °C, ammonia decomposes, and ammonia is also toxic.

In periodical processes, solid absorbents were proposed, (thermochemical refrigeration), especially calcium chloride-ammonia, but the properties of many other working pairs have been studied (Buffington 1933, Eggers-Lura *et al.* 1975, Dueñas 2001, Le Pierrés *et al.* 2007). Examples are the methylamine–water solutions, (Roberson *et al.* 1966), the lithium bromide–water pair mainly employed in air conditioning systems, and sodium thiocyanate in liquid ammonia (Sargent and Beckman 1968, SLGC 1961, Blytas and Daniels 1962). Further examples include combinations of halogenerated hydrocarbons (Zellhoeffer, 1973 Albright *et al.* 1960, 1962, Mastrangelo 1959, Thieme and Albright 1961, Eiseman 1959), alcohol in salt mixtures (Akers *et al.* 1965), and SO_2 in polar organic solvents (Albright *et al.* 1963).

Other substances such as calcium chloride-water, calcium chloride and its hydrates mixed with water, ammonia, and calcium chloride ammines, ammonia and zinc-sulfate ammines, and lithium nitrate as solvent and ammonia as solute, are some examples of the different working substances proposed for absorption refrigeration cycles.

The success of the lithium bromide–water pair in large air conditioning machines fostered a number of very important investigations on the requirement for residential applications, particularly concerning the addition of a third organic or inorganic substance, aiming at modification of the crystallization characteristics of the basic binary system. Examples include the following substances: cesium bromide and ethylene glycol, with high viscosity and then very high pumping requirements, lithium chloride with marginal improvement (Weill 1960), lithium thiocyanate with potential use, zinc bromide (Aronson 1969), lithium iodide with/ without glycols (Hensel and Harlowe1972), with corrosion and high viscosity problems. In order to avoid the corrosion problem, some inhibitors have been proposed: molybdate, zirconate, silicate, titanate, and chromate salts (Weil and Ellington 1956).

In order to reduce the working pressure of ammonia, its replacement by monomethylamine has been proposed, in monomethylamine-water solutions (Pilatowsky *et al.* 2001, Romero *et al.* 2005, Blytas and Daniels 1962), monomethylamine with thiocyanates.

In spite of the large number of refrigerant-sorbent combinations available, only a few pairs are widely used industrial and commercially, mainly the ammonia–water and lithium bromide–water pairs.

5.4.2.1 The Ammonia–Water System

The ammonia–water system is one of the oldest pairs employed for industrial and commercial absorption air conditioning and refrigeration. The ammonia as refrigerant in aqueous solutions has been much used in ARS because of its favorable thermodynamic characteristics, *i.e.*, large negative deviation from Raoult's law, which states that the partial pressures of the different components of an ideal binary mixture are proportional to *their mole fractions* in the liquid and vapor phases, additionally has a low molecular weight and hence large heat of vaporization. However, because of the high affinity between ammonia and water, although their boiling points are spread apart by about 133 °C, some water is always vaporized from generator with the ammonia vapor, and a rectifier system is required. Water is an adequate sorbent of ammonia due to its availability, non-toxicity, and low cost.

5.4.2.2 Lithium Bromide–Water System

Lithium bromide–water is the pair that has been widely used in the operation of air conditioning systems, due basically to the normal freezing point of the water. The principal advantages are: (1) the use of water as refrigerant, which has a very high heat of vaporization, (2) the sorbent is non-volatile, (3) the system operates at low pressure and hence the required pumping power is low, and (4) the material is non-toxic and non-flammable. However, the main disadvantage of using water as refrigerant is that a water cooled condenser is required to attain temperatures corresponding to air conditioning; further the temperatures corresponding to refrigeration applications cannot be reached even with a water cooled condenser. This is because lithium bromide is not sufficiently soluble in water to permit the absorber to be air cooled. This pair has a corrosive effect on the construction materials, and, moreover, during its operation there is the problem of crystallization and hence it is unsuitable for prolonged utilization.

5.4.3 Absorption Refrigeration Cycles

Absorption refrigeration can be used in closed and open cycles for continuous and intermittent operations. This cycle can be considered similar to the vapor compression cycle with the compressor being replaced by a generator-absorber assembly. An absorption system consists of a generator, a condenser, an evaporator and an absorber, and additionally a pump and an expansion valve.

5.4.3.1 The Intermittent Absorption Cycle

According to Figure 5.6, the concentrated solution contained in the generator (G), is heated, causing separation of the refrigerant vapor, and then the vapor is separated from the absorbent to obtain the purest possible refrigerant in the rectifier (R). The pure refrigerant continues toward the condenser (C), where it is liquefied and stored in a recipient (RC). In the intermittent cycle, the period of generation-condensation of the refrigerant and the period of evaporation–absorption occur at different times.

The diluted solution that remains in the generator is then allowed to cool down until it reaches the pressure and the temperature necessary for the refrigerant to be

Figure 5.6 Intermittent
absorption refrigeration cycle

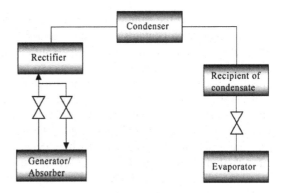

absorbed by the diluted solution. Once this has been achieved, the refrigerant
liquid is sent through an expansion valve toward the evaporator (E), where its
vaporization is carried out at low pressure, producing a low temperature. The
vapor refrigerant returns to the generator and now has the absorber function and
then it is re-absorbed to begin a new cycle.

The fundamental difference among absorption refrigeration cycles in intermittent and continuous operation is that in the intermittent cycle, the generation–
condensation process and evaporation–absorption process are carried out at different times. The intermittent cycle can be developed to constant pressure or constant
temperature.

5.4.3.2 The Continuous Absorption Cycle

As shown in Figure 5.7, the vapor refrigerant enters the absorber at low pressure,
where it is absorbed by the absorbent solution.

The solution (refrigerant/absorbent) that leaves the absorber contains a high
concentration of refrigerant (concentrated solution). This solution is pumped from
the absorber to a generator at high pressure. The concentrated solution enters to
the generator at high pressure and low temperature where it is heated, this elevates

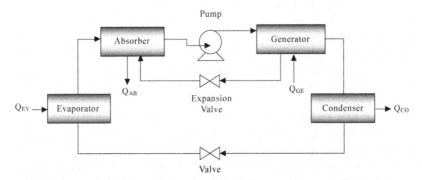

Figure 5.7 Continuous absorption refrigeration cycle

the temperature of the solution to produce saturated vapor. The solution resulting after the generation contains a low concentration of refrigerant (diluted solution). This solution returns to the absorber, passing through an expansion valve, which has the function to cause a fall of pressure to be able to maintain a difference of pressures between the generator and the absorber.

The refrigerant vapor leaves the generator at high pressure and at high temperature and enters into the condenser where its temperature is reduced, producing its liquefaction. The liquid refrigerant passes through an expansion valve where the pressure is abruptly reduced until reaching the evaporation pressure. In the evaporator, the liquid refrigerant is evaporated at low pressure by withdrawing heat from the surroundings (space or body to be cooled), thus cooling it.

The superheated refrigerant vapor leaves the evaporator at low pressure and returns to the absorber where it is re-absorbed by the diluted solution, thus completing the cycle. Condensation and absorption are the exothermic processes, where heat is dissipated by coolant media such as water or ambient air.

The COP of the continuous absorption cycle is defined as

$$\text{COP} = \frac{\text{heat absorbed to vaporize the refrigerant}}{\text{total heat supplied} + \text{pumping work}} = \frac{Q_{EV}}{Q_{GE} + W_P} \qquad (5.19)$$

where Q_{EV} and Q_{GE} are the evaporation and the generation heats of the refrigerant, and W_P the work given at the pump.

5.4.3.3 Modified Continuous Absorption Refrigeration Cycle

The performance of the basic absorption cycle can be improved by recovering the heat losses. The regenerated absorbent solution leaving the generator must be cooled to the exit temperature of the absorber before it leaves the generator, and at the same time the concentrated solution leaving the absorber must be heated to a temperature close to the generation temperature.

The modified continuous absorption refrigeration cycle (refined), incorporates a liquid heat exchanger for transferring the thermal energy of the diluted solution (from generator to absorber) to the concentrated solution (from absorber to generator), saving a substantial amount of energy that would otherwise be wasted and reducing the thermal energy requirements in the generator and the cooling requirements in the absorber.

The sensible heat of the cold vapor leaving the evaporator can be used in order to cool the liquid refrigerant supplied to the expansion valve. This heat exchanger liquid-vapor is known as sub-cooler and increases the refrigerating capacity extracted in the evaporator.

The solution heat exchanger and liquid sub-cooler are not essential components for the cycle operation, but they allow saving energy and making the operation of the system more efficient, increasing the COP.

The main disadvantage of this system is the fact that some absorbents are volatile as water. When the evaporated refrigerant leaves the generator it can to contain some amount of absorbent vapor, which is undesirable because after conden-sation the absorbent can freeze along the pipe. Moreover, when the refrigerant liquid water enters the evaporator it can elevate the evaporation temperature.

In this case, it is necessary to integrate another component that separates the largest amount possible of the absorbent vapor from refrigerant vapor. For this process a rectifier is used, which allows to carry out this operation, offering the transfer heat and mass areas necessary for the partial condensation of the absor-bent vapor.

The operating rectifier temperature should be higher than the condenser tem-perature in order to avoid possible condensation of the refrigerant vapor in the rectifier. The pure condensed absorbent is put in contact with the refrigerant vapor leaving the generator, forming a more concentrated solution than the one formed in the absorber. However, this amount is very small and it does not affect the op-erating conditions of the cycle.

This solution returns to the generator. This process can reduce the amount of absorbent vapor leaving from the generator until the concentration required is achieved. Another approach would be to consider that at the exit of the rectifier the vapor refrigerant concentration is close to 100%, which represents an ideal case. Figure 5.8 represents a modified continuous absorption refrigeration cycle.

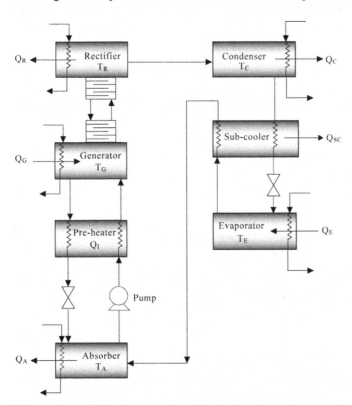

Figure 5.8 Modified continuous absorption refrigeration cycle

5.4.3.4 Absorption Refrigeration Cycle for Air Conditioning

In this absorption refrigeration cycle, the refrigerant is water, which is absorbed in lithium bromide solutions. Because the refrigerant is water, this system does not allow the production of cold at temperatures lower than 0 °C. However, this system is designed for the air conditioning of space. Figure 5.9 schematically shows this type of refrigerating machine. This cooling system is constituted of a group of elements that operate at high pressure: the condenser and the generator, and another group that operates at low pressure: the evaporator and the absorber.

In this process the condensed refrigerant is irrigated onto the tubular external surface of the evaporator, where a flow of water is circulated by means pump (P_1). This water is cooled as it releases the heat required to evaporate the refrigerant. The heat released by the condensation and absorption processes is removed by cooling water.

The vapor formed is absorbed by the solution flowing over the absorber tubes (pump 2). The resulting dilute absorbent solution is transported by the pump (P_3) of the low pressure section toward the high pressure section, through the heat exchanger (HE), where it takes heat from the concentrated absorbent solution, before introducing it into the generator (4). Thus, water vapor is produced and liquefied in the condenser, then evaporated and absorbed, and the thermodynamic cycle is completed.

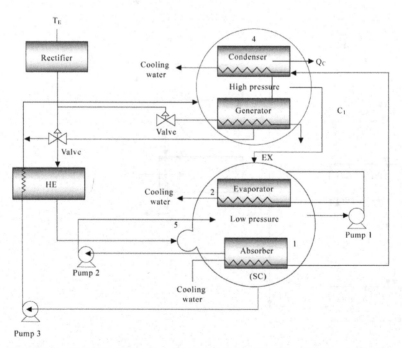

Figure 5.9 Absorption lithium bromide—water system for air conditioning

The liquid formed returns by the circuit (C_1) and the expansion valve toward the evaporator. The concentrated solution (CS) returns toward the low pressure area, cooling down in the heat exchanger (HE). The system feeds in point 5, through the pump (P_2) with an intermediate concentration solution between the diluted and concentrated solutions. The regulator (R) controls the cooling of the water temperature by means of a thermal element (T_E), the flow of the fluid of heating by means of an automatic valve VA, and the flow of the solutions by means of a three-way valve (V_{3w}).

5.5 Advanced Cycles

Various designs of absorption refrigeration cycles have been proposed in order to improve their performance, for example, half, double, tripe, and quadruple effects, insertion of internal exchangers for heat recovery (absorption), and combined system such as vapor absorption–compression cycles, absorption–resorption cycles, dual absorption refrigeration cycles, ejector–absorption refrigeration, diffusion–absorption cycles, and osmotic–membrane absorption cycles (Srikhirin *et al.* 2001). The more interesting cycles correspond to multieffect advanced cycles, which allow improvement of the COP in a significant way.

5.5.1 *Multieffect Absorption Refrigeration Cycles*

The main objective of a higher effect cycle is to increase system performance when high temperature heat source is available. The multieffect cycle (MEC) has to be configured in such a way that heat rejected from a high temperature stage is used as heat input in a low temperature stage for generation of an additional cooling effect in the low-temperature stage.

In this cycle, high temperature heat from an external source is supplied to the first-effect generator. The vapor refrigerant generated is condensed at high pressure in the second-effect generator. The heat rejected is used to produce additional refrigerant vapor from the solution coming from the first-effect generator. This configuration is considered as a series flow double-effect absorption system.

A double effect absorption system is considered as a combination of two single effect absorption systems whose value is COPS. For one unit of heat input from the external source, the cooling effect produced from the refrigerant generated from the first-effect-generator is $1 \times$ COPS. For any single-effect absorption system, it may be assumed that the heat rejected is approximately equal to the cooling capacity obtained. A double effect absorption system has a COP of 0.96, when the corresponding single-effect has 0.6. Theoretical studies of a double-effect absorption system have been provided for working fluids (Kaushik and Chandra 1985, Garimella and Cristensen 1992) Figure 5.10 shows a double-effect absorption using $LiBr$-H_2O.

If LiBr-H$_2$O is replaced with NH$_3$–H$_2$O, the maximum pressure in the first-effect generator will be extremely high. Figure 5.11 shows a double-effect absorption system using NH$_3$–H$_2$O. This system can be considered as a combination of two separate single-effect cycles. The evaporator and the condenser of both cycles are integrated together as a single unit. Thus, there are only two pressure levels, and the maximum pressure can be limited to an acceptable level. The heat from an external source is supplied to generator II. In this case, there is no problem of crystallization in the absorber. Hence, absorber II can be operated at high temperature and rejects heat to generator I. This configuration is considered as a parallel flow double-effect absorption system.

Triple and quadruple-effect absorption cycles have been analyzed (Devault and Marsala 1990, Grossman *et al.* 1995). In the triple-effect, the cycle operates at four

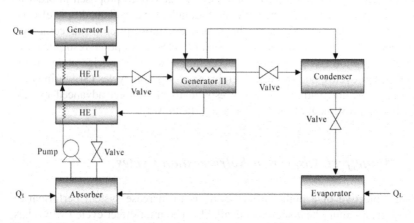

Figure 5.10 A double-effect absorption refrigeration cycle (lithium bromide–water)

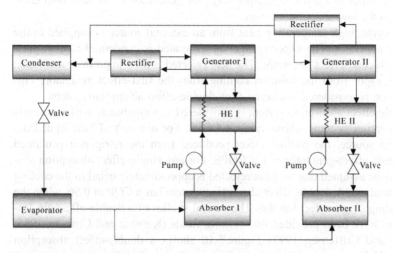

Figure 5.11 Double-effect absorption operating with two pressure levels

pressure levels. Heat of condensation from the higher-pressure stage is used for refrigerant separation in the lower-pressure stage.

However, an improvement of COP is not directly linked to the increment of the number of effects; therefore, the double-effect cycle is the one that is available commercially (Ziegler *et al*. 1993).

5.5.2 Absorption Refrigeration Cycles with a Generator/Absorber/Heat Exchanger

GAX stands for generator/absorber/heat exchanger or sometimes is called DAHX, which stands for resorber/absorber/ heat exchanger. Higher performance can be achieved with a single-effect absorption system. The simplified configuration is shown schematically in Figure 5.12. The term GAX refers to the inserting a heat exchanger (X), between the generator (G) and the absorber (A). In some cases this is known as DAHX, which represents a heat exchanger (X) between a desorber (D) and an absorber (A). High efficiency can be reached with a single-effect cycle. According to the double-effect absorption system in parallel flow, mentioned previously, the system consists of two cycles of a single effect, working in parallel form.

An absorber and a generator may be considered as a counter-flow-heat exchanger as shown in Figure 5.12. At the absorber, diluted solution from the generator and vapor refrigerant from evaporator enter at top section. Heat produced during the absorption process must be rejected in order to maintain the ability to absorb the refrigerant vapor. At the top section of the absorber, heat is rejected at a high temperature. In the lower section, the solution further absorbs the vapor re-

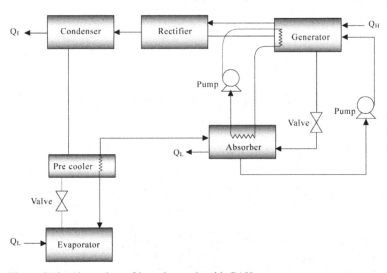

Figure 5.12 Absorption refrigeration cycle with GAX

frigerant while cooling down by rejecting heat to the surroundings. At the generator, concentrated solution from absorber enters at the top section.

In this section, the refrigerant is dried out from the solution as it is heated by using the heat rejected from the top section of the absorber. At the lower section of the generator, the solution is further dried as it is heated by the external source.

Referring to Figure 5.12, there is an additional secondary fluid, which is used for transferring heat between the absorber and the generator. Therefore, a single-effect absorption system can provide a COP as high as that for the two-stage double-effect absorption system using GAX. This system has been studied by Staicovici (1995), Potnis *et al.* (1997), and Priedeman and Christensen (1999).

5.5.3 Absorption Refrigeration Cycle with Absorber-heat-recovery

The use of a solution heat exchanger improves the COP. Concentrated solution from absorber can be preheated before entering the generator by transferring heat from the hot solution coming from the generator. By introducing absorber-heat-recovery, the temperature of concentrated solution can be further increased (Figure 5.13).

Similar to the GAX system, the absorber is divided into two sections. Heat is rejected at a different temperature. The lower temperature section rejects heat to the surroundings. However, the higher temperature section is used to preheat concentrated solution, as shown in Figure 5.13. Thus, the heat input to the generator is reduced, causing the COP to increase. This system was studied theoretically by using various working fluids, for example, water–ammonia and lithium nitrate–ammonia (Kandlikar 1982, Kaushik and Kumar 1987).

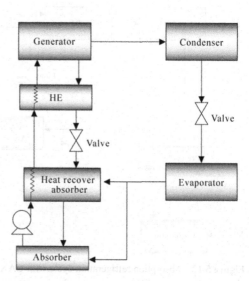

Figure 5.13 Advanced cycle with absorber heat recovery

5.6 Adsorption Refrigeration System

Adsorption heat pumps are solid/vapor systems, very similar in concept to thermochemical systems, apart from the basic difference that the operation between the solid and the gas is an adsorption process. The thermal effect is due to the heat of adsorption. The main component is a container in which the adsorption and desorption of refrigerant vapor takes place. In this case, the solid is now an adsorbent and the vapor refrigerant is an adsorbate.

This recipient is connected to a condenser and an evaporator. As in all sorption systems, there is a relationship of phases that governs the phenomena of adsorption, heat transfer, and pressure, and operation temperature. Adsorption refrigeration is a thermal driven system (TDRS) which can be powered by any kind of thermal energy resource such as waste industrial heat and solar energy.

ARC uses a solid adsorbent to adsorb and to desorbe a refrigerant fluid in order to obtain a cooling effect. The main components are: a solid adsorbent, a condenser, an expansion valve, and an evaporator. The adsorption refrigeration system (ARS) follows the classical thermodynamical intermittent sorption refrigeration single stage cycle.

The adsorbent desorbs refrigerant vapor when it is heated to drive it to the condenser, where it is cooled and liquefied. The refrigerant condensate then expands to a lower pressure through an expansion valve and is transferred to the evaporator, which exchanges heat with the space to be conditioned, allowing its vaporization. When the adsorption conditions are established in the solid bed, the refrigerant vapor from the evaporator is re-introduced to the adsorber and is adsorbed to complete the cycle.

One adsorption refrigerator is advantageous because of its simple structure and low initial cost. The adsorption cycle can only provide intermittent cooling and can be used in the cases when continuous cooling or higher cooling capacity are not required; also, it has lower system performance in terms of COP and specific cooling power (SCP), defined as the ratio between the production of cold and the cycle time *per* unit of adsorbent weight. SCP reflects the size of the system, where high SCP values indicate compactness of the system (Meunier 1998).

As can be seen in Figure 5.14, there are technical problems in the transport of the adsorbent (or solid compound), which influence the intermittent character of the operation of the cycle. The process, in general, consists of a tank where the adsorption is carried out between a solid compound and an adsorbent (activated coal, aluminum oxide, silica-gel, *etc.*) and a refrigerant vapor; this gas is the fluid of work of the cycle. The removed heat corresponds to adsorption heat.

This system operates on the principle that the reversible adsorption between the solid and the gas is subject to the principles of the existent balance among both phases; this reversibility depends on the adsorbent temperature and the vapor pressure of the refrigerant. To obtain a continuous and stable cooling effect in the ARS, generally two or more adsorbers are used. In the two-adsorber cycle (two-bed adsorption), one is heated during the desorption period and the other adsorber is cooled during the adsorption period; then the complete cycle is known as "cycle

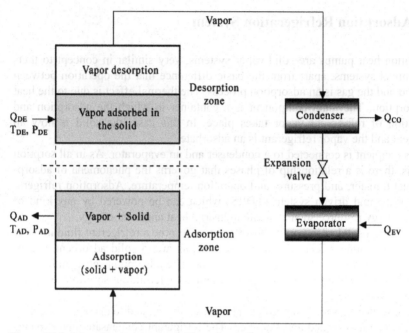

Figure 5.14 Intermittent adsorption refrigeration cycle

time". The heating and cooling periods are reversed when the adsorbers reach the desired upper and lower temperature limits of the refrigeration cycle, which depend on the selection of the refrigerant fluid and the adsorbant. Chillers based on two-bed adsorption have been commercialized in Japan, China, and USA, using silica gel–water as the working pair (Yong and Wang 2007).

5.6.1 Adsorbent/Adsorbate Working Pair

There are several adsorbent–adsorbate working pairs for solid adsorption system, such as molecular sieves, for example, zeolite, silica gel, alumina, active carbon, and some metal salts. Most new adsorbents are based on modification of existing adsorbent material by consolidating, composite, and adding metal material into the adsorbent to improve the heat and mass transfer and increase the adsorption capacity, so that the system's energy performance can be improved and system size can be reduced.

References

Aronson D (1969) Absorption refrigeration system. US Patent 3,478,530
Albright LF, Doddy TC, Buclez PC *et al* (1960) Solubility of refrigerants 11,21 and 22 in organic solvents containing an oxygen atom. ASHRAE Trans 66:423

Albright LF, Shannon PT, Terrier F *et al* (1962) Solubility of chlorofluoro-methanes in non-volatile polar organic solvents. AIChE J 8(5):668

Albright LF, Shannon PT, Yu SN *et al* (1963) Solubility of sulphur dioxide in polar organic solvents. A Chem Symp Ser 59,44:66

Akers JE, Squires RG, Albright LF (1965) An evaluation of alcohol-salt mixtures as absorption refrigeration solutions. ASHARAE Trans 71:14

Andrews DH (1964) Refrigerant-absorbent pairs for absorption refrigeration machines. Washington DC American Gas Association

Buffington RM (1933) Absorption refrigeration with solid absorbents. Refrigeration Eng 26:137

Buffington RM (1949) Quality requirements for absorbent refrigerant combinations. Refrig Eng 57:343–349

Blytas GC, Daniels F (1962) Concentrated solutions of NaSCN in liquid ammonia: solublity, density, vapour pressure, viscosity. Thermal conductance heat of solution and heat capacity. J Amer Chem Soc 84:1075

Cardwell DSL (1971) From Watt to Clausius: the rise of thermodynamics in the early industrial age. Heinemann, London

Carnot S (1824) Reflexions sur la puissance motrice du feu et sur les machines propres á développer cette puissance. Bachelier Libraire, Paris

Clausius R (1850) Über die bewegende Kraft der Wärme. Part I, Part II. Annalen der Physik 79 368–397, 500–524 (1851), see English translation: On the moving force of heat, and the laws regarding the nature of heat itself which are deducible therefrom. Phil Mag 2:1–21, 102–119

Devault RC, Marsala J (1990) Ammonia-water triple effect absorption cycle. ASHRAE Trans 96:676–82

Dueñas I, Pilatowsky I, Romero RJ *et al* (2001) A dynamic study of the thermal behaviour of solar thermochemical refrigerator: barium chloride-ammonia for ice production. Sol Ener Mat Sol C 70:401–413

Eding HJ, Brady AP (1961) Refrigerant-absorbent systems. SRI Project S/3372 Final Report, Stanford Research Institute

Eggers-Lura A, Nielsen P, Stubkier BA, Worse-Schmidt P (1975) Potential use of solar powered refrigeration by an intermittent solid absorption system. Technical University of Denmark

Garimella S, Christensen RN (1992) Cycle description and performance simulation of a gas-fired hydronically coupled double-effect absorption heat pump system. ASE 28, Recent research in heat pump design. ASME 7–14

Grossman G, Zaltash A, Adcock PW *et al* (1995) Simulating a 4-effect absorption chiller. ASHRAE Jun 45–53

Hainsworth WR (1944) Refrigerants and absorbents. Part I and Part II. Refrig Eng 48:97–100

Hensel WE, Harlowe IW (1972) Compositions for absorption refrigeration system. US Patent 3,643,455

Institut International du Froid (1958) Régles pour machines frigorifiques (Kältemaschinenregeln). German Cold Association (translated by Feniger, CK, Paris)

Kandlikar SG (1982) A new absorber heat recovery cycle to improve COP of aqua-ammonia absorption refrigeration system. ASHRAE Trans 88:141–158

Kaushik SC, Chandra S (1985) Computer modelling and parametric study of a double-effect generation absorption refrigeration cycle. Energy Conversion. Manag 25(1):9–14

Kaushik SC, Kumar RA (1987) A comparative study of an absorber heat recovery cycle for solar refrigeration using NH_3-refrigerant with liquid/solid absorbents. Energy Res 11:123–132

Le Pierrés N, Mazet N, Stitou D (2007) Modeling and performances of a deep-freezing process using low-grade solar heat. Energy 32(2):154–164

Macriss RA (1976) Selecting refrigerant absorbent fluid system for solar energy utilization. ASHRAE Trans 82(1): 975–988

Mansoori GA, Patel V (1979) Thermodynamic basis for the choice of working fluids for solar absorption cooling systems. Solar Energy 22(6):483–491

Macriss RA, Gutraj JM, Zawacki TS (1988) Absorption fluid data survey. Final report on worldwide data. ORLN/sub/8447989/3, Institute of Gas Technology

Mastrangelo SVR (1959) Solubility of some chlorofluorohydrocarbons in tetraethylene glycol dimetil ether. ASHRAE J 10:64

Meunier F (1998) Solid sorption heat powered cycles for cooling and heat pumping application. Appl Therm Eng 18:715–729

Niebergall W (1959) Sorptions Kältemaschinen. Handbuch der Kältetechnik, vol VII. Springer, Berlin

Pilatowsky I, Rivera W, Romero RJ (2001) Thermodynamic analysis of monomethylamie-water solutions in a single-stage solar absorption refrigeration cycle at low generator temperatures. Sol Energ Mat Sol C 70:287–300

Potnis SV, Gomezplata A, Papar RA et al (1997) GAX component simulation and validation. ASHRAE Trans 103:444–453

Priedeman DK, Christensen RN (1999) GAX absorption cycle design process. ASHRAE Trans 105(1):769–779

Raldow W (1982) New working pair for absorption processes. In: Workshop Proceedings, Berlin. Swedish Council for Building Research

Roberson JP, Lee CY, Squires RG and Albright LF (1966) Vapor pressure of ammonia and monomethylamine in solutions for absorption refrigeration system, ASHRAE Trans., 72, Part Y, 198–208.

Romero RJ, Guillen L, Pilatowsky I (2005) Monomethylamine-water vapour absorption refrigeration system. Applied Thermal Engineering 25:867–879

Rush WF, Macriss RA, Weil SA (1967) A new fluid system for absorption refrigeration. Fourth International Congress of Heating and Air Conditioning, Paris

Sargent SL, Beckman WA (1968) Theoretical performance of an Ammonia-NaSCN intermittent absorption refrigeration cycle. Sol Energy 12:137

Srikhirin P, Aphornratana S, Chungpaibulpatana S (2001) A review of absorption refrigeration technologies. Renew Sust Energ Rev 5(4):343–372

Staicovici MD (1995) Polybranched regenerative GAX cooling cycles. Int J Refrig 18(5): 318–329

Swedish Council for Building Research (1982) New working pairs for absorption processes. In: Raldow W (ed) Workshop Proceedings, Berlin

Thomson W (Lord Kelvin) (1851) Dynamical theory of heat. Royal Soc Edin 3:48–52

Tyagi KP, Rao KS (1984) Choice of absorbent-refrigerant mixtures. Energy Res 8:361–368

Weil SA (1960) Thermodynamic properties of lithium chloride, lithium bromide–water system. Report IGT, Project No. S/153, Institute of Gas Technology

Weil SA, Ellington RT (1956) Corrosion inhibition of lithium bromide–water cooling systems. Project ZB-29, Institute of Gas Technology

Yong L, Wang RZ (2007) Desorption refrigeration: a survey of novel technologies. Recent Patents on Engineering 1:1–21

Zellhoeffer GF (1937) Solubility of halogenated hydrocarbon refrigerants in organic solvents. Ind Eng Chem 29:548

Chapter 6
Cogeneration Fuel Cells –
Air Conditioning Systems

6.1 Introduction

As already discussed in Chapter 1, energy is a finite resource and its rational use implies an increase in energy efficiency. The electric generation efficiency is always less than 100% due to resistive, transmission and distribution losses, which can be quantified as heat sent to the environment. This waste heat determines the quantity of energy that can be used by other systems in order to improve the process efficiency.

A comfortable climate is required in offices, houses, and sport facilities. Therefore, cooling and heating requirements increase proportionally to the world population growth. According to the above idea, the suitability to satisfy these comfort requirements is in the use of waste energy to activate air conditioning machines. In this sense, fuel cells are a very attractive source of waste heat and therefore this possibility should be explored.

This chapter describes the mathematical models and assumptions considered for coupling a fuel cell with air conditioning systems, specifically absorption heat pumps (AHP). Three absorption refrigeration systems, operating with different mixtures or working fluids for its operation, are simulated. The results for each case studied and a comparison in terms of efficiency is presented next.

6.2 Considerations for Cogeneration Systems Based on Fuel Cells and Sorption Air Conditioning

The theory of integration of a proton exchange membrane fuel cells (PEMFC) with thermal systems was presented recently by Pilatowsky *et al.* (2007). Both the fuel cell and the absorption system described in that work were already experimentally tested for each particular use.

PEMFC have been experimentally evaluated to estimate the quantity of electric power produced from hydrogenesis reactions. In these experiments, the relationship between the electric and thermal power obtained has been investigated. Ex-

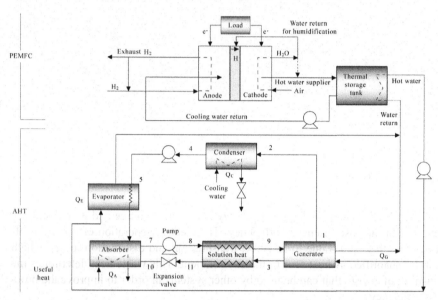

Figure 6.1 A PEM fuel cell coupled with an absorption refrigeration system

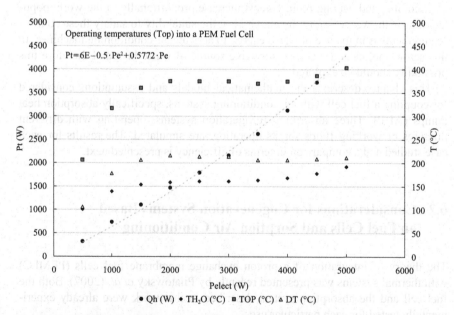

Figure 6.2 Relationship of thermal and electric energy in a laboratory cell

perimental tests have given enough evidence to confirm that for a PEMFC with a nominal capacity up to 5 kW, a maximum thermal power of 4.5 kW can be obtained. In general, a relationship of 10:9 is obtained.

The diagram of a PEMFC fuel cell and its coupling with an absorption facility is shown in Figure 6.1. The nonlinear relation between electric and thermal power in a laboratory fuel cell is presented in Figure 6.2. This relation is used for simulation purposed for the cases presented in Section 6.3.

Absorption systems have had a discreet participation in the air conditioned market (IEA, 2001). The main obstacle for their diffusion is the complexity of its operation and the fact that most cooling systems have been implemented for particular conditions.

In the following, the considerations of the coupling of absorption systems with PEM fuel cells is described. This could be an attractive application to allow the introduction of these systems in the market.

6.2.1 Coupling of Technologies

In order to simulate the integration of fuel cells and air conditioning systems, the following assumptions were considered:

1. The systems operate under steady state conditions.
2. The heat exchangers have finite heat transfer areas.
3. The non-electric energy of the fuel cell is dissipated entirety in the form of heat.
4. The absorption heat pump uses the thermal energy dissipated by the fuel cell.
5. A differential temperature is considered between the outlet dissipation fluid temperature and the entrance generation temperature to the AHP.

Figure 6.3 shows a block diagram representing the integration of a fuel cell and an absorption air conditioning system. Electric energy is obtained from the fuel cell and the thermal energy Q_T is obtained as byproduct. The energy dissipated by the fuel cell Q_T is used by the absorption system, which requires heat to operate, as explained in Chapters 4 and 5. The main energy (98 %) that makes the AHP work

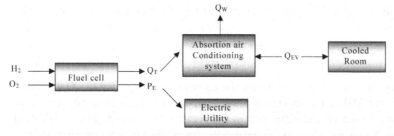

Figure 6.3 Block diagram of the coupling of a fuel cell and an absorption system

is in form of heat (Romero *et al.* 2007). The result of this coupling is electric power P_e and a cooling effect Q_{EV}.

It is important to consider that the system operation depends on the energy given by the stechiometric conversion of hydrogen and oxygen in water, plus an enthalpies difference. This enthalpies difference is the quantity Q_T given by the fuel cell.

6.2.2 Concepts of Efficiency

The coupling of systems has the advantage of increasing their efficiency (Rivera *et al.* 2007). For sorption systems, the coefficient of performance COP is a measure of the system efficiency, defined as the ratio between the useful energy and the energy supplied to the system. This COP is broadly used in heat pumps, whether they are operated by compression or absorption. COP is defined as the ratio of the sum of the useful energy ΣQ_U to the input energy ΣQ_{IN}:

$$COP = \frac{\Sigma Q_U}{\Sigma Q_{IN}} \tag{6.1}$$

For the case of absorption heat pump, the evaporator power Q_{EV}, considered useful, is the numerator in Eq. (6.1) and the thermal energy supplied to the generator Q_T is the denominator. In this way, a cooling COP can be defined for air conditioning, as follows:

$$COP_{AC} = \frac{Q_{EV}}{Q_T} \tag{6.2}$$

When COP is equal to 1, it indicates that a cooling capacity will be obtained, practically similar to the quantity of thermal energy that is entered into the system.

The conversion efficiency of a PEMFC is associated to the quantity of energy that can contribute to a recycled flow of hydrogen to saturate the membrane of the fuel cell and can obtain a reaction with better conversion. The details of this system were given in Chapters 2 and 3.

For the case under study in this chapter, the energy dissipated by the fuel cell is totally used. The electric power produced by the fuel cell is known through the conversion efficiency. The fuel cell efficiency is defined (Pilatowsky *et al.* 2007) as:

$$\Delta H = \frac{W_{elect,\,max}}{\eta_{elect}} \tag{6.3}$$

This way it is possible to determine the quantity of energy that the fuel cell can transfer to the AHP, which occurs at lower temperature than that in the hydrogenesis reaction. However, the difference in temperature does not diminish the power exchanged, because the heat exchange process takes place without phase change.

A liquid circuit is used in order to dissipate the heat for the correct functioning of the fuel cell. A fluid enters at initial temperature T_i and exits at a higher temperature T_f. This difference is proportional to the power that is transferred between the fuel cell and the AHP. This power has been determined as:

$$Q_T = (m_W)(C_P)(T_F - T_i)$$

(6.4)

Where C_P is the heat capacity and depends on the fuel cell operating temperature. This value remains practically constant during the operation of the system. However, the generator operating temperature in the AHP has a lower value than T_f at the fuel cell outlet. The entrance to the system is the exit cooling fluid fuel cell temperature $T_{GE,in}$ and the exit is $T_{GE,out}$. T_f is higher than $T_{GE,in}$, according to a given coupling temperature difference ΔT_C, which is assumed for simulation. The value $T_{GE,out}$ can be calculated in the following way

$$T_{GE,out} = \frac{Q}{(m_w)(C_P)} + T_F - \Delta T_C$$

(6.5)

6.3 Modeling of Cogeneration Systems Using Fuel Cells. Promising Applications

Chapter 4 shows that several mixtures can be used for sorption systems operating nowadays. The modeling of the coupling of a PEMFC and absorption systems using three known mixtures is presented next. The selection of the mixtures under study is based on the following approaches:

1. The mixture is commercially available.
2. The mixture can operate in a wide temperature range suitable for the fuel cell conditions.
3. The thermodynamics properties and energy transfer coefficients of the mixture are reported previously (Rivera *et al.* 1998, Romero *et al.* 2005).
4. The conditions explained in Chapters 2 and 5, regarding thermodynamic equilibrium in all components, are completed for modeling purposes.
5. The energy losses to the atmosphere are negligible.
6. The pressure losses are negligible in the heat exchange process.
7. The generator mass flow rate entering the AHP and leaving the fuel cell is constant.
8. The fuel cell power is constant under operating conditions.
9. The cooling effect or cooling thermal load is constant.
10. The intermediate temperature for depends on the environmental temperature.

Keeping in mind these considerations, the following calculation for the system coupling can be used.

The electric power produced by the fuel cell is known and constant:

$$P_e + Q_T = k(\eta_{elect}) \qquad (6.6)$$

The fuel cell efficiency is known and, therefore, the available power for the AHP system can be estimated, considering a nonlinear behavior:

$$P_e = \frac{\left[(3.8819)(1000P_T)^{0.8584}\right]}{1000} \qquad (6.7)$$

This approach coincides reasonably well with previous work (Pilatowsky *et al.* 2007, EG&G 2000).

The air conditioning Coefficient of performance, COP_{AC} depends on the internal flows and enthalpies of the AHP. COP_{AC} is calculated from the operating conditions in the generator and evaporator. Therefore, it is possible to estimate the cooling power as:

$$Q_{EV} = (COP_{AC})(Q_T) \qquad (6.8)$$

For cogeneration systems, a new efficiency concept is used, which relates the electric and thermal power obtained with the total energy supplied. The overall efficiency of the system, called efficiency of cogeneration (EOC), is the ratio between the sum of the useful energy and the input energy used by the fuel cell:

$$EOC = \frac{P_e + P_T}{P_{H2}} \qquad (6.9)$$

This EOC is expressed as a function of the electric power in the following way:

$$EOC = \frac{Q_{EV} + P_e}{(P_e)(\eta_{elect})} \qquad (6.10)$$

where the energy potential of hydrogen can be evaluated as a function of conversion efficiency. Having these parameters, the complete modeling with different working fluids of the AHP can be carried out. This allows the comparison and selection, according to the operating and environmental conditions, of the best coupling option.

6.3.1 Operation Conditions

The assumed operating conditions to carry out a comparison among the proposed systems are the following:

1. The cooling power should be constant at 1.0 kW.
2. The environmental temperature must be in the range from 20 °C to 30 °C.
3. The condensation process of the AHP is carried out at those environmental temperatures, with cooling water.

4. The dissipation fluid temperature range operating in the fuel cell is from 60 °C to 80 °C. Therefore, the entering fluid temperature range considering the DT_C, could be from 55 °C to 75 °C.
5. The fuel cell operates with a minimum power of 1.5 kW and a maximum of 2.8 kW.
6. The fuel cell efficiency can be considered constant at 0.57 in a temperature range of 60 °C–80 °C.
7. The electrical power (P_E) was correlated as $3.8819\,Q_{GE} + 0.8584$ (Eq. 6.7).
8. The COP values of the AHP are between 0.36 and 0.92, depending on the operating conditions. This range is obtained from the simulation for these operating conditions.
9. The evaporator temperature range is 4 °C–12 °C.
10. The heat exchanger effectiveness in the AHP is considered as 0.8 (Rivera *et al.* 1998).

6.3.2   Modeling of a Cogeneration System Using an Absorption Air Conditioning System with Water–Lithium Bromide as Working Fluid

The water–lithium bromide mixture, as mentioned in previous chapters, is the reference working mixture for absorption systems, operating for air conditioning and it is commercially available. The simulation results for the coupling between a fuel cell and an absorption air conditioning system (AACS) using this working pair are presented in Figure 6.4. It shows the resulting efficiency of cogeneration EOC as a function of the generation temperature T_{GE} for several evaporation tem-

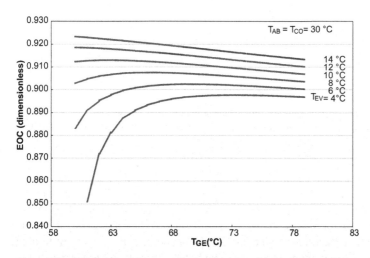

Figure 6.4 Efficiency of Cogeneration against T_{GE} for the cogeneration system operating the AACS with the water–lithium bromide mixture for different T_{EV} at $T_{AB–CO} = 30$ °C

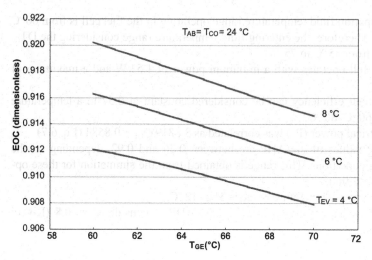

Figure 6.5 Efficiency of Cogeneration against T_{GE} for the cogeneration system operating the AACS with the water–lithium bromide mixture for different T_{EV} at $T_{AB-CO} = 24\,°C$

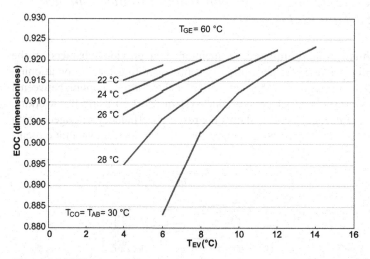

Figure 6.6 Efficiency of Cogeneration EOC against evaporation temperature T_{EV} at different T_{AB-CO} for the cogeneration system operating the AACS with the water–lithium bromide mixture. $T_{GE} = 60\,°C$

peratures T_{EV} and a fixed absorption-condensation temperature T_{AB-CO}. The behavior of the efficiency of cogeneration against T_{GE}, shows that EOC increases with an increment of T_{EV}, which can reach values of $4\,°C$. Also, for higher values of T_{EV} the EOC decreases for higher values of T_{GE}. However, for lower T_{EV} the EOC increases to reach a maximum value and then it decreases. It is important to note that for any T_{EV}, a maximum EOC is reached at T_{GE} less than $70\,°C$, which fits with the operating temperature of PEMFC.

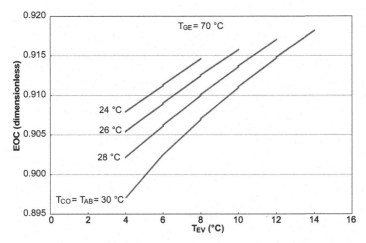

Figure 6.7 Efficiency of Cogeneration EOC against T_{EV} at different T_{AB-CO} for the cogeneration system operating the AACS with the water–lithium bromide mixture. $T_{GE} = 70\,°C$

The same variables are presented in Figure 6.5 with a different T_{AB-CO}. A similar decreasing behavior of EOC is observed with decreasing T_{EV} and increasing T_{GE}. In this case, the system cannot operate at T_{EV} higher than 8 °C, because of crystallization problems. The EOC varied from 0.908 to 0.92.

Figure 6.6 presents the EOC against T_{EV} for different absorption–condensation temperatures and a fixed T_{GE}. The EOC increases with an increment in T_{EV} and a decrease of T_{AB-CO}. For most of the of absorption–condensation temperature values, the T_{EV} range is 6 °C–14 °C; however, for $T_{AB-CO} = 22\,°C$ the system can only operate at T_{EV} from 4 to 6 °C, but with higher EOC, which varied from 0.915–0.925.

In Figure 6.7, the same variables are presented with a different T_{GE}. The behavior of the EOC against T_{EV} is similar to that shown in Figure 6.6, but the system cannot operate at $T_{AB-CO} = 22\,°C$ due to crystallization conditions at this temperature. The EOC values are similar than those obtained with $T_{GE} = 60\,°C$.

6.3.3 Modeling of a Cogeneration System Using an Absorption Air Conditioning System with a Water–Carrol™ as Working Fluid

The water–Carrol™ mixture was patented by Carrier© for absorption systems in the past (Reimann and Biermann 1984). It presented a very useful characteristic for the AHP: the solubility of the mixture allows a wider operating temperature range. The increase of viscosity caused by the additives does not affect the AHP performance with high temperatures in the generator and the absorber (Romero *et al.* 2007). It is a potential mixture for AHP with four temperature levels, where the higher temperatures correspond to the generator and the absorber.

The simulation results for the coupling of a PEMFC with an AHP operating with the water–Carrol™ mixture are given in Figure 6.8, where the EOC is plotted against T_{GE} for different evaporation temperatures T_{EV} and a fixed T_{AB-CO}. The efficiency of cogeneration EOC increases with an increment of T_{EV} and decreases with an increment of T_{GE}. However, for lower T_{EV} the EOC increases up to an optimum value. Comparing these results with those obtained for the water–lithium

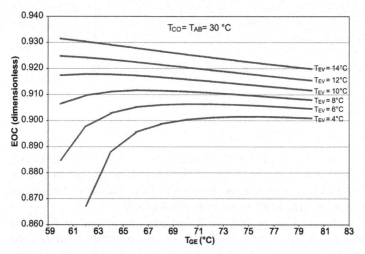

Figure 6.8   Efficiency of Cogeneration EOC against T_{GE} for the cogeneration system operating the AACS with the water–Carrol™ mixture for different T_{EV} at $T_{AB-CO} = 30\,°C$

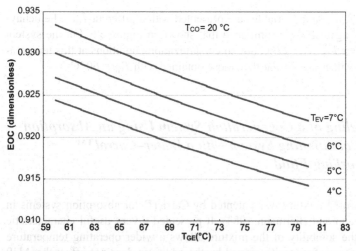

Figure 6.9   Efficiency of Cogeneration EOC against T_{GE} for the cogeneration system operating the AACS with the water–Carrol™ mixture for different T_{EV} at $T_{AB-CO} = 20\,°C$

bromide mixture (Figure 6.4), it can be seen that the EOC is slightly higher when using water–Carrol™ mixture, with values over 0.93.

The same variables, EOC vs. T_{GE}, are plotted in Figure 6.9, but with T_{AB-CO} fixed to 20 °C. The behavior of the EOC is similar to that shown for the case of water–lithium bromide mixture (Figure 6.5); however, as for the previous case, the EOC values are a slightly higher with the present mixture.

Figure 6.10 presents the EOC against T_{EV} for different absorption–condensation temperatures and a fixed T_{GE}. The EOC increases with an increment of T_{EV} and the

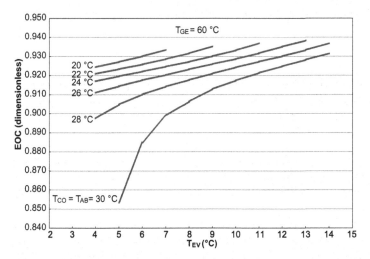

Figure 6.10 Efficiency of Cogeneration EOC against T_{EV} for the cogeneration system operating the AACS with the water–Carrol™ mixture for different T_{AB-CO} at $T_{GE} = 60$ °C

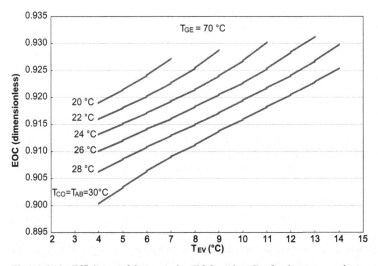

Figure 6.11 Efficiency of Cogeneration EOC against T_{EV} for the cogeneration system operating the AACS with the water–Carrol™ mixture for different T_{AB-CO} at $T_{GE} = 70$ °C

decrease of T_{AB-CO}. For higher T_{AB-CO} values the system can operate at T_{EV} in a range from 5 °C to 14 °C. However, at T_{AB-CO} = 20 °C the system can only operate at T_{EV} from 4 °C to 7 °C, but reaching the highest values of EOC which varied from 0.924–0.934.

In Figure 6.11 the EOC against T_{VE} is presented for a different T_{GE}. As in the previous case, the EOC increases with higher values of T_{EV} and with a decreases in T_{AB-CO}. Comparing the EOC values with those shown in Figure 6.7 for the water–lithium bromide mixture, it can be appreciated that with the water–Carrol™ mixture the system can operate at absorption–condensation temperatures of 20 °C and 22 °C, while the water–lithium bromide mixture cannot operate at those temperatures.

6.3.4 Modeling of a Cogeneration System Using an Absorption Air Conditioning System with Monomethylamine–Water as Working Fluid

The mixture monomethylamine–water (MMAW) is a potential candidate for applications with moderate generation temperatures. The coupling of this system was analyzed in a recent work (Pilatowsky 2007, Romero *et al.* 2007) with successful results.

The coupling of a fuel cell with an absorption heat pump operating with this mixture presents a variation with regard to the previous cases. For this case, a rectification is required after the generation, with a slight demand for heat. Both generation and rectification processes occur at moderate temperatures.

In Figure 6.12, the simulation results for the coupling between a fuel cell and an a refrigeration system using monomethylamine–water are presented. The resulting cogeneration efficiency as a function of T_{GE} is plotted for several evaporation temperatures and a fixed absorption–condensation temperature T_{AB-CO}. A decreasing behavior of the EOC with the increase in T_{GE} can be observed, similar to the previous cases studied. But the EOC decreases more rapidly with an increase of $T_{GE,}$ varying from 0.75 to 0.88, which are lower than those obtained using the previous mixtures.

Figure 6.13 depicts the results for EOC vs. T_{GE} but with T_{AB-CO} = 20 °C. A decrease in EOC is observed with an increment of T_{GE}. Comparing the EOC values with those presented in Figures 6.5 and 6.9, it is noticeable that the system using the MMAW mixture can operate over a wider range of T_{EV} with similar EOC values.

The results for EOC against T_{EV} for different absorption–condensation temperatures are presented for a fixed T_{GE} in Figure 6.14. The increase in EOC is observed with an increment of T_{EV}. At lower T_{EV} the difference in EOC are considerable, with the change of condensation–absorption temperatures T_{AB-CO}. However, at higher T_{EV} the EOC practically does not vary with variation of T_{AB-CO}. In Figure 6.15 the EOC against T_{VE} is presented for a different T_{GE}. Similarly to the previous case, for lower T_{EV} there is considerable a difference in EOC when T_{AB-CO} is varied; but at higher T_{EV} the EOC is barely affected by the change in T_{EV}. Comparing the EOC obtained in this case with those obtained with the previous mixtures (Figures 6.7 and 6.11) it can be observed that the maximum EOC is 0.86, while for the other mixtures it is higher than 0.91.

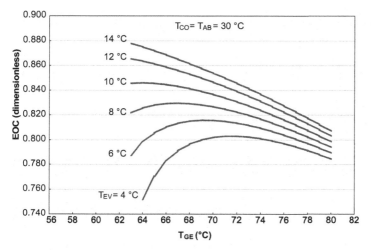

Figure 6.12 Efficiency of Cogeneration EOC against T_{GE} for the cogeneration system operating the AACS with the monomethylamine–water mixture for different T_{EV} at $T_{AB-CO} = 30\,°C$

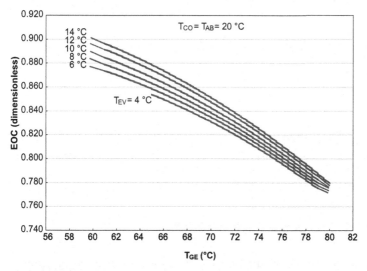

Figure 6.13 Efficiency of Cogeneration EOC against T_{GE} for the cogeneration system operating the AACS with the monomethylamine–water mixture for different T_{EV} at $T_{AB-CO} = 20\,°C$

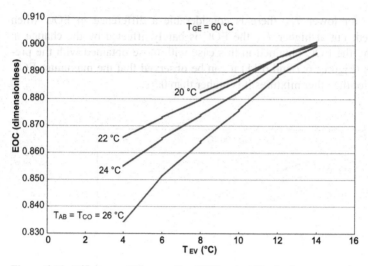

Figure 6.14 Efficiency of Cogeneration EOC against T_{EV} for the cogeneration system operating the AACS with the monomethylamine–water mixture for different T_{AB-CO} at $T_{GE} = 60\,°C$

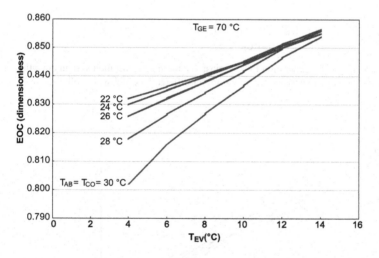

Figure 6.15 Efficiency of Cogeneration EOC against T_{EV} for the cogeneration system operating the AACS with the monomethylamine–water mixture for different T_{AB-CO} at $T_{GE} = 70\,°C$

6.4 Modeling of Trigeneration Systems

Trigeneration can be defined as the production of three types of energy from a single source. A particular case of trigeneration is the integration of a fuel cell with an absorption heat pump to obtain simultaneously electric power, heat and cooling (Figure 6.16). To carry out this trigeneration system, AHP types I and II

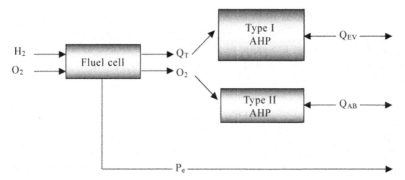

Figure 6.16 Trigeneration system consisting of a fuel cell with two AHPs for simultaneous electric power, heating, and cooling

described in previous chapters are considered for the integration. The main characteristic that makes them different is that the evaporation temperature is lower for type I. For type I, the lowest temperature is obtained in the evaporator. In type II, it is obtained in the condenser.

The coupling for a trigeneration system requires the following assumptions:

1. The fuel cell transforms the energy from the hydrogenesis to electric and thermal forms.
2. Two types of AHP systems are integrated with a fuel cell.
3. The dissipation process occurs without phase change.
4. The dissipation fluid temperature range operating in the fuel cell could be from 60 °C to 80 °C.
5. The intermediate temperature for AHP type I depends on the environmental temperature.
6. The entering fluid temperature range considers the DT_C.
7. The energy losses to the environment and pressure drop inside the AHP are negligible.
8. The "economizer component" is a heat exchanger into the AHP with effectiveness from 0.7 to 0.8.

The efficiency of the trigeneration for this system is defined in the following way:

$$EOT = \frac{Q_{EV} + Q_{AB} + P_e}{P_{H2}} \qquad (6.11)$$

This can also be expressed as a function of the nominal fuel cell electric power as

$$EOT = \frac{Q_{EV} + Q_{AB} + P_e}{P_e \cdot h_{elect}} \qquad (6.12)$$

Where Q_{EV} and Q_{AB} are the quantity of energy exchanged with the environment for cooling and heating, respectively. The power Q_{EV} is obtained from an AHP type I and Q_{AB} it is obtained from an AHP type II. The energy supplied to both AHPs is the total quantity of thermal energy exchanged by the fuel cell.

For the case of cooling, the value of the COP_{T1} is calculated as the ratio between the evaporation power, divided by thermal energy (Q_1) given by the fuel cell to this AHP. This value is calculated for an AHP type 1

$$COP_{T1} = \frac{Q_{EV}}{Q_1} \tag{6.13}$$

For heating, the value of the COP_{T2} is calculated as the relationship of the thermal energy given to AHP of type II. The total energy given to this AHP is part of the thermal energy that the fuel cell rejects.

$$COP_{T2} = \frac{Q_{AB}}{Q_2} \tag{6.14}$$

The energy given to both AHP is the energy rejected by the fuel cell. That is to say

$$Q_T = Q_1 + Q_2 \tag{6.15}$$

This way, the value of the trigeneration efficiency EOT can be defined as

$$EOT = \frac{[COP_{T1}Q_1 + COP_{T2} + P_e]}{P_e \eta_{elect}} \tag{6.16}$$

This efficiency of trigeneration EOT will depend on the variables of each coupling, using the Equation 6.15 for the convergence of the system. The values obtained will always be higher than 1. This occurs because there is a part of energy from the surroundings which is used without cost, represented by the two first terms in the numerator.

The simulation results for this system, considering the coupling of a fuel cell that operates with the above mentioned conditions, are the following:

$P_e = 2\,kW$
$T_f = 80\,°C$
$\eta_{elect} = 0.43$
$Q_T = 1.27\,kW$
$Q_{EV} = 0.500\,kW$ @ $T = 4\,°C$
$COP_{T1} = 0.961$
$Q_1 = 0.52\,kW$
AHP Type I mixture: MMAW
$Q_{AB} = 0.562\,kW$ @ $T = 106\,°C$
$COP_{T2} = 0.464$
$Q_2 = 0.75\,kW$
AHP Type II mixes: water–Carrol™
EOT = 3.3

These values indicate that the system recovers energy from the fuel cell for both heating and cooling effects. The dissipation heat is used to obtain cooling effect at 4 °C and heating effect at 106 °C.

The alternative cogeneration systems based on fuel cells can be integrated with diverse heat pumps systems, as types I and II, which can lead to useful for applications in domestic systems or pre-thermal treatments.

The potential for the coupling with AHP type I depends on the temperatures and powers required. The most promising systems are those operating with MMAW and H_2O–LiBr. For the case of the integration with AHP type II, the most promising mixture is water–Carrol™. The main challenge of coupling with either type I or type II is to assure that the fuel cell operates with the highest efficiency possible to produce electric power. To take advantage of the cogeneration system, it is important to keep the fuel cell temperatures constant and heat loads should be similar to the electric power, to avoid that electric efficiency diminishes. The proposal of combined systems is an attractive solution to the lack of energy resources in the near future.

6.5 Conclusion

The modeling of different integration system using a fuel cell with absorption systems, working with three different mixtures, was presented. The results show that the lowest evaporation temperature was 4 °C obtained for all mixtures studied. The main parameter, the cogeneration efficiency EOC, provides an useful comparison of the systems. The EOC will always be higher than the COP of each absorption air conditioning system operating individually. The EOC for the system operating with monomethylamine–water is lower compared with those obtained for water–lithium bromide, and water–Carrol™, being higher for the case of water-lithium bromide. The most promising absorption air conditioning system integrated with an fuel cells is water–Carrol™, which can operate at generation temperatures lower than 70 °C, with 20 °C for condensation temperature.

References

EG&G (2000) Science Applications International, Corporation, US Department of Energy. In: Fuel cell handbook, 5th edn. Parsons

Morrillón GD (2002) Introducción a los sistemas pasivos de enfriamiento. In: RIRAAS – sistemas de enfriamiento aplicados a la vivienda. CD 3(4):1–21

Pilatowsky I, Romero RJ, Isaza CA, Gamboa SA, Rivera W, Sebastian PJ, Moreira J (2007) Simulation of an air conditioning absorption refrigeration system co-generation process combining to proton exchange membrane fuel cell. Int J Hydrogen Energ 32:3174–3182

Reimann R, Biermann WJ (1984) Development of a single family absorption chiller for use in solar heating and cooling system. Phase III Final Report. Prepared for the US Department of Energy under contract EG-77-C-03-1587, Carrier Corporation

Rivera C, Pilatowsky I, Méndez E, Rivera W (2007) Experimental study of a thermo-chemical refrigerator using the barium chloride-ammonia reaction. Hydrogen Energ 32:3154–3158

Rivera W, Cardoso MJ, Romero RJ (1998) Theoretical comparison of single stage and advanced absorption heat transformers operating with water/lithium bromide and water/Carrol mixtures. Int J Energ Res 22:427–442

Romero RJ, Rodríguez Martínez A, Casillas González E (2007) Laboratory instrumentation and object oriented design for working fluid control in an "absorption heat pump" using water/Carrol. In: Sobh T (ed) Innovations and advanced techniques in computer and information sciences and engineering, 1st edn. Springer, Netherlands

Romero RJ, Guillén L, Pilatowsky I (2005) Monomethylamine-water vapour absorption refrigeration system. Appl Therm Eng 25:867–879

Sebastian PJ, Gamboa SA (2007) Local grid connected integrated solar-hydrogen-fuel cell systems in Mexico. Int J Hydrogen Energ 32(15):3109

Chapter 7
Potential Applications
in Demonstration Projects

7.1 Introduction

Fuel cells are devices where electrical energy is produced by redox electrochemical reactions of hydrogen or enriched hydrogen fuels and oxygen, or simply air. Powerful fuel cells or stacks have emerged as potential replacements for the internal combustion engine in automobile vehicles, because the use of fuel cells implies clean and efficient use of energy, and the unique by-products are water and heat when pure hydrogen is used as fuel.

Fuel cells can find applications in transport, stationary power generation, supply of heat and electricity in buildings, in space missions for providing electricity and water, and military applications for providing electric power to electronic devices. The available energy of fuel cells is greater than the energy stored in primary or secondary batteries. This has led to a great interest in the development of fuel cells for many applications.

The use of fuel cell based technologies can be a significant support for lowering the energy consumption involved in the demand of the actual societies, where the amount of energy utilized by communities, is an indicator of a country's development level. It is possible to consider the use of hydrogen in fuel cells from any natural source or hydrogen enriched fuels, and nowadays there are many efforts and examples of utilizing fuel cells performing with hydrogen enriched fuels. The prognosis in this area shows the feasibility of using fuel cells not just in mobile or portable applications but in industrial and residential applications. The conjunction of fuel cells performing with thermal engines in CHP systems is expected to be a useful and easier way to obtain energy at the lowest energetic and economic reliability.

PEMFC is one of the most investigated and utilized fuel cells due to its adequate design and operation mode. PEMFC was first used in aerospace applications as an auxiliary power source in the Gemini Space Vehicle program in the early 1960s. Some performance problems of PEMFCs, such as very low power density and short lifetime (less than 400 h use) in space vehicles was the reason to propose a different fuel cell in alkaline medium (AFC), and NASA's Apollo and space shuttle programs have been using AFC for supplying energy in space flights.

PEMFCs had another opportunity to be considered as an alternative for supplying energy when the Nafion membrane was developed and the ability of the PEMFC for maximizing its power density and the lifetime was really enhanced. This was probably the great chance for supplying energy at room temperature and obviously it was clearly thought to be used in terrestrial applications. In the near future the concept of wireless systems will be completely applied when laptops, mobile telephones, and other portable devices will have a PEMFC instead of batteries or grid connectors. These advantages were an incentive for the fast development of PEMFC in the last ten years as part of the US program for promoting a new generation of ultra-low emission vehicles in 2004. Batteries were chosen as a complement of hybrid power systems in order to save energy and its disposal in critical energy demand like start up and acceleration of engines.

The use and scope of the investment in fuel cell research, development, and demonstration projects in the latter half of the 20th century and the first part of the 21st century have been focused on the power generation/cogeneration systems for improving the conversion efficiencies of conventional fossil energy sources (coal, petroleum, natural gas, and nuclear energy) into electricity. With this strategy it has been possible to establish a new mechanism to properly use existing energy and promote energy conservation.

Major interest at present is focused on fuel cells for stationary power generation and for portable and automotive purposes. In stationary power fuel cells it is possible to use different types of fuel cells according to the fuel disposal and criteria for designing the power system.

7.2 A New Era in Energy Revolution: Applications of Fuel Cells

Fuel cells are electrochemical systems designed to convert the chemical energy of hydrogen and oxygen reactants into electricity, heat, and water. The application of fuel cells has been shown in many feasible and demonstrative projects around the world, in stationary, mobile (transportation), portable, and military uses.

Maybe the most known projects are related to the automotive industry and isolated and grid interconnected projects for electrical energy supplies in buildings, hospitals and houses. Fuel cells are also important in the military industry as power supply of special devices, principally for electronic systems. The use of fuel cells in CHP systems has opened the possibility to obtain energy efficiencies around 100 %. Some examples of the use of fuel cells in novel projects to produce electricity are presented in the following sections, and the possibility to take advantage of the heat and water produced from the fuel cells used in cogeneration systems can be a good opportunity to diminish the disadvantages of this new technology obtaining an efficiency close to 100 % of the total energy involved in the CHP systems for electrical power generation.

Fuel cells are not batteries, but they have similar properties such as the ability to transform chemical energy of reactants into electricity. Basically fuel cells require hydrogen enriched fuels and oxygen to work in a continuous fashion; oxygen is easy to take from the air, but hydrogen is not in a free form on earth, and when free it is volatile, hazardous, and difficult to store and handle. Hydrogen can be extracted from several different fuels.

Natural gas, gasoline, diesel, and propane contain massive amounts of hydrogen that can be used in fuel cells. Renewable energy and fuels also offer hydrogen; these fuels can be ethanol, methanol, landfill gas, bio-gas, and methane. Water is the most abundant molecule on Earth. Through electrolysis, oxygen and hydrogen can be separated for direct use. There are other novel forms of hydrogen fuel, including peanut shells, algae, and sodium borohydride.

Currently, hydrogen is very costly, which depends mostly on the delivery system of the hydrogen. Hydrogen can be pumped through pipelines. Hydrogen can also be drawn from the above fuels and carried on board cars, like gas in combustion engines. The best alternative would be for fueling stations to be powered by solar, wind, or hydroelectric power, and to use that electricity to separate hydrogen from water using electrolysis.

Nowadays, there are many uses where fuel cells are involved as primary or secondary electrical power systems. All major automobile companies are working to commercialize fuel cell cars. Fuel cells power buses, boats, trains, planes, scooters, forklifts, even bicycles. There are fuel cell-powered vending machines, vacuum cleaners, and highway road signs. Miniature fuel cells for cellular phones, laptop computers, and portable electronics are on their way to market. Hospitals, credit card centers, police stations, and banks all use fuel cells to provide power to their facilities. Wastewater treatment plants and landfills use fuel cells to convert the methane gas that they produce into electricity. Telecommunication companies install fuel cells at cell phone, radio, and television re-transmission towers. The possibilities are endless but some examples of these applications are given in the following four sections.

7.2.1 Stationary Applications

There are more than 2000 fuel cell systems around the world, and they have been installed in hospitals, nursing homes, hotels, office buildings, schools, utility power plants, *etc.* They have been connected to an electrical grid to provide supplemental power and backup assurance for critical areas, or installed as a grid-independent generator for on-site service in areas that are inaccessible by a conventional electrical grid.

The efficiency (fuel to electricity conversion ratio) of fuel cell power generation systems has achieved around 40 % and it is higher than the efficiency obtained from hydrocarbon based machines. Since fuel cells do not have mobile parts, the operation is silent and they reduce noise pollution as well as air pollution. When the fuel cell is placed close to the operating point, its waste heat can be reused for energy cogeneration purposes. In large-scale building systems, these fuel cell cogeneration systems can reduce the use of conventional electricity by up to 40 % and the efficiency can reach 85 %.

In telecommunications with the use and demand of computers, internet, and communication networks increasing, there is also a need to support the performance of telecommunications in more reliable electric power plants than the available and conventional electrical grid. Under specific situations, fuel cells have proven to be as reliable as electric power plants. Fuel cells can replace batteries to provide electric power from 1–5 kW telecommunication sites without noise or emissions, and are durable, providing power in places that are inaccessible or difficult to reach at some period of the year. Such systems would be used to provide primary or backup power for telecom switch nodes, cell towers, and other electronic systems that would benefit from on-site, direct DC power supply.

Other important and very recent use of fuel cells is related to landfills and wastewater treatment plants, where fuel cells operate as a demonstrative and valid technology for reducing emissions and generating electric power from the methane gas they produce.

It is thought that fuel cells have the potential to produce electricity for homes, businesses, institutions, and industry through stationary power plants. The size in electric demand is from 1 kW to several MW.

Stationary power applications of fuel cells will require the development of low-cost, reliable, and efficient power inverter and grid interface technology. Power inversion is required to convert the direct current power ordinarily produced by the fuel cell stack into the alternating current on the utility grid. Also, control technology is needed to achieve reliable and cost-effective operation of fuel cells and to produce high-quality power. The balance of plant considerations such as pumps, valves, piping, controls, and power electronics requires advancement in reliability, cost, and optimization for fuel cell applications.

It is possible to think that the use of fuel cells in stationary applications will be, necessarily, in a scheme of distributed generation. This means a small-scale electric generation located close and enough to the demanded electric load. Sometimes this is also referred to as "on-site power" or "distributed power", which includes not only generation but also storage. Distributed generation is often in contrast to central generation. In the case of central generation, power is generated in a large plant and electricity is transmitted over transmission and distribution lines (collectively referred to as the power grid) to buildings where the power is consumed, but normally the consumption zones are far from the generation plants, implying energy losses that can increase the final cost of the electricity.

In the case of distributed generation, the potential exists to provide generation at the building where the power is consumed and the excess of energy can be supplied to the power grid in an interactive model. Another key advantage of distributed generation is that in addition to the higher efficiency in electricity, the heat produced by the fuel cell can be utilized by the consumer sites; it is associated to the concept of combined heating and power systems.

Some examples of companies having fuel cell projects for stationary applications are: Avista Labs, Ebara Ballard, Fuel Cell Energy, Fuel Cell Technologies, GE Microgen, General Motors, H Power, Idatech, Matsushita Electric Industrial Co Ltd, Nuvera, Plug Power, Inc., Proton Energy Systems, Sanyo Electric Co., Siemens Westinghouse Power Corp., Toyota Motor Corporation, and UTC fuel cells.

7.2.2 Mobile and Transportation Applications

All major automotive companies have a fuel cell vehicle even under development or as testing prototype. Commercialization is a little further down the line (some automakers say 2012, others even later), but every demonstration helps bring that date closer.

In the last 4 years, more than 50 fuel cell buses have been demonstrated all around the world. Fuel cells are highly efficient, so even if the hydrogen is produced from fossil fuels, fuel cell buses can reduce CO_2 emissions, and the emissions can be zero if the hydrogen is produced from a renewable source, improving air quality. Because the fuel cell system is so much quieter than a diesel engine, fuel cell buses significantly reduce noise pollution as well.

In spite of their small size, many scooters are pollution power houses. Gas-powered scooters, especially those with two-stroke engines, produce tailpipe emissions at a rate disproportionate to their small size. These two-stroke scooters produce almost as much particulate matter as and significantly more hydrocarbons and carbon monoxide than a heavy diesel truck. Fuel cell scooters running on hydrogen will eliminate emissions – in India and Asia where much of the population uses them – this is a great application for fuel cells.

Fuel cell forklifts have the potential to effectively lower total logistics cost since they require minimal refilling and significantly less maintenance than electric forklifts, whose batteries must be periodically charged, refilled with water, and replaced. Due to the frequent starting and stopping during use, electric forklifts also experience numerous interruptions in current input and output; fuel cells ensure constant power delivery and performance, eliminating the reduction in voltage output that occurs as batteries discharge.

Nowadays, heavy duty trucks are equipped with a large number of electrical appliances from heaters and air conditioners to computers, televisions, stereos, even refrigerators and microwaves. To power these devices while the truck is parked, drivers often must idle the engine. The Department of Energy (DOE) has estimated the annual fuel and maintenance costs of idling a heavy duty truck at over US $1,800 and that using fuel cell APUs in Class 8 trucks would save 670 million gallons of diesel fuel *per* year and 4.64 million tons of CO_2 emission *per* year.

Fuel cells are being developed for mining locomotives since they produce no emissions. An international consortium is developing the world's largest fuel cell vehicle, a 109 metric-ton, 1 MW locomotive for military and commercial railway applications.

Fuel cells are attractive options for aviation since they produce zero or low emissions and make barely any noise. The military is especially interested in this application because of the low noise, low thermal signature,and ability to attain high altitude. Companies like Boeing are heavily involved in developing a fuel cell plane.

For each liter of fuel consumed, the average outboard motor produces 140 times the hydrocarbons produced by the average modern car. Fuel cell engines have higher energy efficiencies than combustion engines and, therefore, offer better range and significantly reduced emissions. Iceland has committed to converting its vast fishing fleet to use fuel cells to provide auxiliary power by 2015 and, eventually, to provide primary power in its boats.

For transportation applications of fuel cells, one of the most important developments involves fuel handling and fuel processing. For example, the proton exchange membrane fuel cell (PEMFC), which is considered to be the primary candidate for transportation propulsion applications, needs a pure, clean hydrogen fuel. Therefore, stringent requirements must be placed on the processing of transportation fuels like gasoline and methanol to eliminate compounds that may poison the cells.

The development of compact, efficient, cost-effective, high purity, hydrogen-producing reformer technology is a key requirement. An alternative strategy to relieve the need for on-board reformation of liquid transportation fuel is the storage and direct use of hydrogen. This approach will require significant advances in the storage of hydrogen using metal hydrides, carbon nanotubes, or cash-worthy, high-pressure hydrogen tank technology. A significant development of the hydrogen supply infrastructure would be needed as well.

Some examples of automotive companies developing mobile applications using fuel cells are: Daimler Chrysler, Ford, General Motors, Honda, Hyundai, Mazda, etc.

Space missions have also used fuel cells for providing electricity and water principally to the crew, and new fuel cells that develop and provide for more ambitious projects in the future are being designed. Independence in energy and water can be crucial to achieve the proposed projects.

7.2.3   Portable Applications

Fuel cells can provide power where no electric grid is available and they are quiet. Thus, using one instead of a loud, polluting generator at a campsite would not only save emissions, but also would not disturb nature or your camping neighbors either.

Portable fuel cells are also being used in emergency backup power situations and military applications. They are much lighter than batteries and last much longer, which is especially important for soldiers carrying heavy equipment in the field.

Fuel cells will change the telecommuting world, powering cellular phones, laptops, and palm pilot hours longer than batteries. Companies have already demonstrated fuel cells that can power cell phones for 30 days without recharging and laptops for 20 h. Other applications for micro fuel cells include pagers, video recorders, portable power tools, and low power remote devices such as hearing aids, smoke detectors, burglar alarms, hotel locks, and meter readers. These miniature fuel cells generally run on methanol, an inexpensive wood alcohol also used in windshield wiper fluid. Some examples of companies involved in the development of fuel cell portable systems are: Ball Aerospace, Motorola, Ballard Power Systems, NEC, Casio, Panasonic, Direct Methanol Fuel Cell Corp., Polyfuel, H Power, Samsung Advanced Institute of Technology, and Hitachi.

7.2.4 Military Applications

Fuel cells may provide power for most types of military equipment, from land and sea transportation to portable handheld devices used in the field. Thus, military applications are expected to become a significant market for fuel cell technology. The efficiency, versatility, extended running time, and noiseless operation make fuel cells extremely well suited for military applications.

Clearly, fuel cells would have many advantages over conventional batteries. As a beginning there would be no need to worry about the logistics of supplying spare batteries. In a similar way, the efficiency of fuel cells for transport would dramatically reduce the amount of fuel required during use. Since the 1980s, the US Navy has used fuel cells for deep marine exploration craft and unmanned submarines.

7.2.5 Combined Heat and Power

Combined heat and power (CHP) systems allow the production of useful heat and electricity from the same utilized source, thus, increasing the overall efficiency through recovery of the heat rejected from the inefficient energy conversion for producing electricity. The use of CHP systems is not a new concept but it is an excellent way to take advantage of using a major portion of the total energy obtained from the fuel cells (heat and electricity principally) to be converted into useful energy. Normally, a typical CHP system has the following components: the engine that drives the generator, the generator that produces electricity, the heat recovery system that collects the waste heat from both the engine water cooling jacket and exhaust gases, the exhaust system to take away the products of combustion, and the control panel to monitor the operation. The complete system is complex and the utilization of any other component depends on the project characteristics.

Micro combined heat and power systems to be used in domestic application fuel cells and cogeneration for office buildings and factories are in the mass production stage. The stationary fuel cell application generates constant electric power and at the same time produces hot air and water from the waste heat. A lower fuel-to-electricity conversion efficiency is accepted (typically 15–20 %), because most of the energy not converted into electricity is released as heat. In terms of exergy, however the process is inefficient and it can improve by maximizing the electricity generated and the electricity is used to drive a heat pump. With some specific fuel cells, the combined efficiencies are around 80–90 %.

7.3 Examples of Combined Heat and Electricity Use from Fuel Cells in Demonstration Projects

7.3.1 Stationary PAFC Cogeneration Systems

Phosphoric acid fuel cells have been developed to a technological degree their commercialization is possible. Tunkey 200 kW plants are available and hundreds have been installed in America, Europe, and Japan. The operating temperature is around 200 °C, and they have shown the capability of supplying hot water as well as electricity, depending on the specific design of the plants. In all cases, the electrical efficiencies have exceeded 40 % (The institution of Engineering and Technology Factfile 2006).

There are two types of fuel cells, which are not treated in this book, with excellent characteristics for producing electricity and heat (solid oxide fuel cells and molten carbonate fuel cells), the unique problem is the high operating temperature.

7.3.2 PEMFC in Mobile Systems

Low and intermediate fuel cells such as phosphoric acid fuel cells and alkaline fuel cells have been used in portable and transportation applications, as well as demonstration vehicles. Some inherent problems like intolerance to carbon dioxide (alkaline fuel cells) and the big size for mobile purposes (phosphoric acid fuel cells) have allowed that low temperature, the solid polymer fuel cell, especially, the proton exchange membrane fuel cell, could emerge as an important fuel cell for portable, mobile, and stationary applications.

The output of a PEMFC can be quickly modified to supply the demanded load, thus making it suitable for mobile purposes where fast start-up is required. When designing systems it is good to consider that fuel cells, in stacks, produce around 1 kWhr *per* liter of hydrogen, equivalent to 0.7 kWhr/kg or 28 kWhr/ft^3.

When it is not possible to store hydrogen as fuel in a portable or mobile device, or the fuel is obtained *via* a reformation process, then the useful system is the alcohol fuel cell; the direct methanol fuel cell is the most studied cell. The alcohol can be obtained from hydrocarbons or by biological methods.

Table 7.1 Typical fuel cell applications

Typical applications	Portable electronic equipment	Automobiles, boats, domestic CHP			Distributed power generation, CHP, buses	
Main advantages	Higher energy density compared to batteries, faster recharging	Potential for zero emissions, higher efficiency			Higher efficiency, less pollution, quiet operation	
Power (W)	1 10 100	1 K	10 K	100 K	1 M	10 M
Application range for FC		AFC			MCFC	
		←	→	←		→
				SOFC		
		←				→
	PEMFC					
	←			→		
			PAFC			
			←	→		

Hydrocarbon fuelled fuel cells have considerable potential for the automotive industry. They avoid the high cost of hydrogen and the low specific energy density of hydrogen when it is compressed; liquid fuels do not only have a much higher specific energy density, but also the benefits of an established distribution system. The range of hydrocarbon fuels that can be used as hydrogen sources includes not only methanol but also natural gas and gasoline. PEMFCs using such fuel sources are currently being analyzed by a number of automobile manufacturers trying to develop zero-emission engines.

As a result of the inherent size, flexibility, and operating principles of fuel cells, it is possible to find a technological range of applications where fuel cells can be used as electric power supplies as well as cogeneration systems. The potential applications of fuel cells can be from few watts to megawatts of electric or combined power.

Table 7.1 shows some typical fuel cell applications (Larminie and Dicks 2000).

The use of fuel cells in automobile applications started in the early 1990s with the efforts made by Toyota, Ford, Daimler Chrysler, etc. Commercial fuel cell powered vehicles are expected by 2010.

7.3.3 CHP Systems with Fuel Cells

The combined use of heat and electricity is the main application in stationary fuel cell projects. The use has been considered for buildings, industrial facilities, or backup generators. Because of the efficiency of fuel cell power systems, which is scalable with size, it is possible to design several hundred kilowatts to low megawatts capacity plants. The most common operation mechanism of the plants implies the use of natural gas in combination with the different types of fuel cell technologies (PEMFC, AFC, PAFC, PCFC, SOFC) (EG&G Service Parsons 2000).

7.3.3.1 Hospital and Autonomous Applications

Phosphoric acid fuel cells have been used in hospital and autonomous applications, where the combined use of electricity and heat has resulted in more efficient processes. Fiji Electric Advanced Technology Co. Ltd, has developed many projects using two 100 kW PAFC commercial models. The first model was used in hospital, hotel, colleges, and office buildings from 1998 to 2000. The second commercial model was used initially in 2001, and there is a report on applications in 2006, where 100 kW PAFCs were used to operate induction centers, sewage disposal plants, hospitals, hotels, exhibition facilities, and office buildings (Oka 2006). Fuel cells have been operating for more than 40,000 h. As general information, in the case of Daido hospital (Japan), a 100 kW PAFC operates continuously for 24 h as a base power source. Two gas engines are operated only in the daytime on weekdays for excessive loads; the recovered heat is used for pre-heating water and air conditioning. With the complete system it is possible to provide around 50 % of the electrical power consumption, 80 % of the hot water supply, and 10 % of the air conditioning needs in the hospital.

A novel application of PEMFC in a CHP system is the case of the energetic autonomy of an underwater glider performance by using CHP from a PEMFC. (Wang *et al.* 2007), have shown the feasibility of using and combining the heat and electricity produced for a miniature PEMFC to enhance the power efficiency in an underwater glider. This system uses the available heat energy for navigational power, and the electricity generated by the PEMFC is used for the control

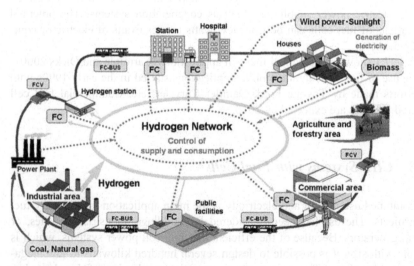

Figure 7.1 The hydrogen society

and electronic system. Experimental results showed an improvement in the thermal engine performance due to the high quality heat from the PEMFC compared with the ocean environmental thermal energy.

Other projects are starting and energy efficiency is increased by using a combined process of the heat and the electricity generated by a fuel cell. At the end, considering the projects where the fuel cells are used in portable, mobile, and stationary CHP applications, it is possible to think that the world, in the near future, is a hydrogen based society as proposed by Murakami (2006), see Figure 7.1. Then, electrical power and thermal energy obtained from FCs will be used properly in a more ordered society.

References

EG&G Service Parsons (2000) Science Applications International Corporation, Fuel cell handbook, 5th edn. National Energy Technology Laboratory

Larminie J, Dicks A (2000) Fuel cell systems explained. Wiley, New York

Murakami Y (2006) The effect of hydrogen on fatigue properties of metals used for fuel cell system. Int J Fracture 138:167–195

Oka Y (2006) Fiji Electric Advanced Technology Co. Ltd, Fuel Cell Seminar Honolulu, HI

The Institution of Engineering and Technology Factfile (2006) www.theiet.org

Wang S, Xie C, Wang Y, Zhang L, Jie W, Hu SJ (2007) Harvesting of PEMF fuel cell heat energy for a thermal engine in an underwater glider. J Power Sources 169:338–346

Chapter 8
Profitability Assessment
of the Cogeneration System

8.1 Introduction

From Chapter 1 it is clear that nowadays electricity and cooling demands are growing considerably worldwide, and that also there is a necessity to provide these by
using more efficient and cleaner technologies, as the proposed in the present book.

A profitability assessment is determinant to decide the implementation and use
of some technologies; because of this in the present chapter, an economic analysis
of a cogeneration system that consists of a PEM fuel cell coupled to an absorption
air conditioning system PEMFC-AACS is presented.

Several studies about the profitability of PEM fuel cells have been made by
Farbir and Gómez (1996), Tsuchiya and Kobayashi (2004), Bar-On et al. (2002),
Zoulias and Lymberopoulos (2007), and Chang et al. (2007). In these studies, the
authors detail in a very precise way which are the elements and parameters determinant that make profitable the use of this technology. Also they show future
scenarios where the PEM fuel cell can be introduced massively in the market.

Because of the above, the present chapter does not show a detailed profitability
assessment of the PEM fuel cells alone; the idea is to analyze the profitability of
the cogeneration system in an integrated way considering that the use of a PEM
fuel cell has already been decided.

When a profitability assessment of a new technology is being carried out, it is
always important to compare the results with more conventional systems in order
to have all the relevant information before deciding whether or not to use the new
technology. Because of this a PEM fuel cell, together with a conventional compression air conditioning system are also analyzed in the present chapter.

In order to understand and to carry out the economic analysis, in this chapter
the elements of profitability assessment are first presented. Then the profitability
assessment is done separately for a PEM fuel cell, a conventional compression air
conditioning system, and an absorption air conditioning system. Finally the profitability of a cogeneration system conformed by the PEM fuel cell coupled to an

absorption air conditioning system PEMFC-AACS is compared with a PEM fuel cell and a compression air conditioning system PEMFC-CACS.

8.2 Elements of Profitability Assessment

8.2.1 Time Value of Money

Time value of money (TVM) is an important concept in economic engineering, since it can be used to compare investment alternatives and to solve problems involving loans, mortgages, leases, savings, and annuities.

TVM refers to the fact that money in hand today is worth more than money that is expected to be received in the future. This means that the value of money changes with time. This change occurs due to a phenomenon called inflation. Inflation is related to the loss of the spending power of money. In developed countries the inflation is usually between 2–5 % annually, while in developing countries is considerably higher, in some cases reaching values higher than 100 %.

8.2.1.1 Future Worth

Future worth refers to the amount of money to which an investment or loan will grow over a finite period of time at a given interest rate. In other words, the future worth F of the money is higher than its present worth P. When money is invested, the interest is related to inflation, but when money is loaned the interest is not only related to the inflation but also to the risk that the money may not be returned.

When simple interest is used, the future value of money of a present value with an interest is

$$F = P(1 + ni) \tag{8.1}$$

Where i is the fractional interest rate *per* period, and n is the number of interest periods, which in general is given in months or years.

Simple interest is computed only on the original amount borrowed or invested, while compound interest is calculated each period on the original amount borrowed plus all unpaid interest accumulated to date. In the present chapter the compound interest is always assumed in the economic analysis.

When compound interest is used F and P are related by the following equation:

$$F = P(1 + i)^n \tag{8.2}$$

which can be rewritten as

$$F = Pf_i \tag{8.3}$$

where

$$f_i = (1 + i)^n \tag{8.4}$$

which is called the compound interest factor.

8.2.1.2 Present Worth

Similarly as with future worth, the present value P of a future sum of money F is given by the equation

$$P = \frac{F}{(1+i)^n} \tag{8.5}$$

or

$$P = Ff_d \tag{8.6}$$

where

$$f_d = \frac{1}{(1+i)^n} \tag{8.7}$$

which is called the discount factor. It is important to note that

$$f_d = \frac{1}{f_1} \tag{8.8}$$

8.2.1.3 Future Worth of an Annuity

An investment can be a single sum of money deposited at the beginning of the first period, annuities, or both. The future worth of a single sum of money was given in Section 8.2.1.1. An annuity is a series of equal payments in equal time periods. Usually, the time period is 1 year, which is why it is called an annuity. However, the time period can be shorter, or even longer. An "annuity due" is an annuity where the payments are made at the beginning of each time period; for an "ordinary annuity", payments are made at the end of the time period. In the present chapter only ordinary annuities are considered.

The future worth of an annuity is then the total amount of money in the future of a series of equal annual payments A made at the end of each year invested at a fractional rate i over a period of n years. It can be shown that the equation that relates the future worth F and the annuity is (Holland *et al.* 1999)

$$F = A\left[\frac{(1+i)^n - 1}{i}\right] \tag{8.9}$$

which can be written as

$$F = \frac{A}{f_{AF}} \tag{8.10}$$

where

$$f_{AF} = \frac{i}{(1+i)^n - 1} \tag{8.11}$$

which is named the annuity future worth factor.

8.2.1.4 Annuity of Future Worth

From the definitions given in the previous section it is clear that the annuity of future worth is the sum of the equal annual payments required at a fractional rate i over a period of n years to build up a future sum of money. From Equation 8.9

$$A = F\left[\frac{i}{(1+i)^n - 1}\right] \tag{8.12}$$

From Equation 8.10

$$A = Ff_{AF} \tag{8.13}$$

The future worth of an annuity is then the total amount of money in the future of a series of equal annual payments A made at the end of each year invested at a fractional rate i over a period of n years

8.2.1.5 Present Worth of an Annuity

Analogous to the future worth, the present worth of an annuity is the present value of equally spaced payments in the future. Combining Equations 8.5 with 8.12, the present worth of an annuity can be estimated as

$$P = A\left[\frac{(1+i)^n - 1}{i(1+i)^n}\right] \tag{8.14}$$

or

$$P = \frac{A}{f_{AP}} \tag{8.15}$$

where

$$f_{AP} = \frac{i(1+i)^n}{(1+i)^n - 1} \tag{8.16}$$

is the present worth of the annuity factor or also the capital recovery factor (CRF).

8.2.1.6 Annuity of Present Worth

The annuity of present worth is the sum of the equal annual payments A required at a fractional rate i over a period of n years to build up a present amount of borrowed money. From (8.14)

$$A = P\left[\frac{i(1+i)^n}{(1+i)^n - 1}\right] \tag{8.17}$$

Combining Equations 8.16 and 8.17 the annuity of the present worth can be rewritten as

$$A = Pf_{AP} \tag{8.18}$$

8.2.2 Annual Costs and Cash Flows

In order to analyze the profitability of one of some projects it is important to know the different cost and cash flows associated to them. Because of this, in the present section the main costs involved in the economic analysis are explained and defined.

The revenue from the annual sales of products A_S minus the total annual cost or expenses required to produce and sell the product A_{TE} is the annual cash income A_{CI}, which excludes any cost for plant depreciation

$$A_{CI} = A_S - A_{TE} \tag{8.19}$$

A_{TE} includes all the expenses necessary to sell a product, which can be directly or indirectly related to the manufacturing of a product. The annual direct manufacturing expenses include materials, operating labor and supervision, operating maintenance, operating supplies, *etc.*, while the annual indirect manufacturing expenses includes rent and insurances, storage, packaging, quality control, and overheads, amongst others.

The net annual cash income A_{NCI} is the annual cash income A_{CI}, minus the annual amount of tax A_T.

$$A_{NCI} = A_{CI} - A_T \tag{8.20}$$

The net annual cash flow A_{CF} is the net annual cash income A_{NCI}, minus the annual expenditure of capital after the plant has been built A_{TC}

$$A_{CF} = A_{NCI} - A_{TC} \tag{8.21}$$

When plant modifications or additions are not required after the plant has been built $A_{TC} = 0$, furthermore $A_{CF} = A_{NCI}$.

8.2.3 Capital Costs

Capital costs are costs incurred on the purchase of land, buildings, construction, and equipment to be used in the production of goods or the rendering of services. According to Holland *et al.* (1999) the total capital costs C_{TC} of a plant or project consists of the fixed capital costs C_{FC} plus the working capital costs C_{WC} plus the cost of land and other non-depreciable costs C_L

$$C_{TC} = C_{FC} + C_{WC} + C_L \tag{8.22}$$

The fixed capital costs C_{FC} include the cost of land, buildings and structures, engineering and construction, and major equipment (*i.e.*, fuel cells, absorption cooling systems, solar collectors, tanks, heat exchangers, auxiliary facilities, installation, instrumentation, piping, insulation, *etc.*).

The working capital costs, C_{WC}, include raw materials for the goods productions, cost of handling of materials and goods, cost of inventory control, insurances, cash emergencies, and additional cash required to operate the process, *etc.*

The fixed capital costs are costs incurred on the purchase of land, buildings, construction, and equipment to be used in the production of goods, while the working capital costs are completely recoverable at any time (theoretically) due to the fact that no tax allowances are made for depreciation. Although the term depreciation has different meanings, in economic engineering it is the decrease in the economic value of the fixed capital costs.

The straight line method is the simplest and most commonly used method to estimate the average annual amount of depreciation (AD). This can be done by the following equation (Holland *et al.* 1999)

$$A_D = \frac{\left(C_{FC} - SV\right)}{n} \tag{8.23}$$

where SV is the scrap value of the plant after a useful life of n years.

8.2.4 Methods for Estimating Profitability

Nowadays there are several methods of profitability assessment; however, only four are the most used for comparing alternative investments: (1) payback period, (2) NPV or NPC, (3) IRR, and (4) EUAC.

8.2.4.1 Payback Period

Although the payback period PBP is not a method for estimating profitability, strictly speaking, because it does not properly account for the time value of money and ignores any benefits that occur after the payback period, this method is widely used due to its simplicity to analyze one investment or to compare different alternatives under the same considerations.

From Section 8.2.2, the net annual cash flow A_{CF} is the net annual cash income A_{NCI}, minus the annual expenditure of capital after the plant has been built A_{TC}

$$A_{CF} = A_{NCI} - A_{TC} \tag{8.24}$$

and the depreciable fixed capital investment is equal to the fixed capital cost C_{FC} minus the SV of the plant S. Then the payback period PBP is the time required for the cumulative net cash flow from the start up of the plant to equal the depreciable fixed capital investment. PBP is the value of n that satisfies

$$\sum_{n=0}^{n=PBP} A_{CF} = C_{FC} - S \tag{8.25}$$

If the annual cash flows A_{CF} can be assumed to be constant over a number of years, the payback period PBP in years can be calculated as

$$PBP = \frac{C_{FC} - SV}{A_{FC}} \tag{8.26}$$

8.2.4.2 Net Present Value and Net Present Cost

Net Present Value

The net present value NPV, also known as net present worth, is one of the most used methods in financial theory and engineering economics to analyze the profitability of an investment or project and also to compare different alternatives. This method in comparison to the payback period PBP is based on the TVM.

NPV is defined as the sum of the annual discounted cash flows over a period of n years. It is the difference between the present value of cash inflows and the present value of cash outflows. In other words, the NPV of an investment or project is the difference between the sum of the discounted cash flows that are expected from the investment and the amount which is initially invested. This parameter expresses the profitability of a project.

As was shown in Section 8.2.1.2, the present value P of a future sum of money F is given by Equations 8.6 and 8.7, which is called the discount factor. Thus the annual discount cash flow A_{DCF} can be obtained from the annual cash flow A_{CF} multiplied for the discount factor

$$A_{DFC} = A_{CF} f_d \tag{8.27}$$

or

$$A_{DCF} = \frac{A_{CF}}{(1+i)^n} \tag{8.28}$$

By definition

$$NPV = \sum A_{DCF} \tag{8.29}$$

Furthermore combining Equations 8.28 and 8.29, the net present value NPV can be estimated as

$$NPV = \sum \frac{A_{CF}}{(1+i)^n} - I_0 \tag{8.30}$$

where A_{CF} can be calculated from Equation 8.21 and I_0 is the investment at the beginning of the project. If NPV < 0 the project should be rejected. If NPV > 0, the project should be accepted from the economic point of view. If NPV = 0, the decision whether to accept or reject the project should be based on different criteria. When two or more projects have been compared, the one yielding the higher NPV should be selected. Although NPV measurement is widely used for making investment decisions the main disadvantage is that it does not account for flexibility/uncertainty after the project decision.

Net Present Cost

In economic engineering there are some kinds of projects where because of their inherent characteristics there are no sales or incomes. In these cases it is common to use the parameter net present cost (NPC), which can be obtained combining Equations 8.21 and 8.30 where the annual cash incomes $A_{NCI} = 0$ and A_{TC} are the annual total costs

$$NPC = \sum \frac{A_{TC}}{(1+i)^n} + I_0 \qquad (8.31)$$

Because, as in this case, most of the amounts are expenditure or costs, the minus sign associated to the costs is changed by the plus sign and the SV of the plant at the end of the project's life should be subtracted from the costs. When two or more projects are being evaluated using this parameter the project with the lowest NPC should be selected.

8.2.4.3 Internal Rate of Return

The internal rate of return method (IRR), also called discounted cash flow rate of return (DCFRR), is the rate of return promised by an investment project over its useful life. It is sometimes simply referred to as the yield of the project. Unlike NPV, which indicates value or magnitude, IRR is an indicator of the efficiency or quality of an investment. IRR is computed by finding the discount rate that equates the present value of a project's cash outflow with the present value of its cash inflow. In other words, IRR is the discount rate that will cause the NPV of a project to be equal to zero.

Furthermore from Equations 8.29 and 8.30 the IRR is the fractional interest that makes the following equation equal to zero:

$$NPV = \sum A_{DCF} = \sum \frac{A_{CF}}{(1+i)^n} - I_0 = 0 \qquad (8.32)$$

The main advantage of IRR over NPV is that the former is independent of the zero or base year chosen. In contrast the NPV varies according to the zero bases chosen. The simplest and most direct approach when the net cash inflow is the same every year is to divide the investment in the project by the expected net annual cash inflow. This computation will yield a factor from which the IRR can be determined.

8.2.4.4 Equivalent Uniform Annual Cost

The EUAC is the uniform annuity equivalent to the discounted total cost. In other words, it is the annualized sum of the equivalent total present value of capital and operating expenditures incurred during the life of the project.

From Equation 8.17 and substituting a single present value P for the NPV of the project the EUAC is

$$EUAC = NPV\left[\frac{i(1+i)^n}{(1+i)^n-1}\right] \tag{8.33}$$

From Equation 8.16 the annuity present worth factor or the CRF is

$$CRF = \frac{i(1+i)^n}{(1+i)^n-1} \tag{8.34}$$

Therefore, from Equations 8.16 and 8.33 the EUAC can be calculated as

$$EUAC = NPV * CRF \tag{8.35}$$

8.3 Profitability Assessment of the Systems

In order to assess the profitability of the proposed cogeneration system, it is important to compare the results with more conventional systems in order to have all the relevant information before deciding whether or not to use the new technology. Because of this, the cogeneration system consisting of a PEM fuel cell coupled to an absorption air conditioning system PEMFC-AACS is compared with a PEM fuel cell and a conventional compression air conditioning system PEMFC-CACS.

The profitability assessment was done separately for each one of the systems involved in the study to provide specific data and typical costs. With this information the NPC and the EUAC were estimated for the systems and the results obtained were taken as reference values. Then the different costs involved were varied within a reasonable range to carry out a sensitivity analysis of the economic parameters. Finally, the profitability assessment was done comparing the NPC and EUAC values for PEMFC-AACS and PEMFC-CACS.

8.3.1 Profitability Assessment of a PEM Fuel Cell

As was shown in Chapter 3, a fuel cell system consists of three main components that work together: a fuel reformer, a fuel cell stack, and the electronic controls and power conversion equipment. A typical PEM fuel cell stack consists of a series of individual cells, which have proton exchange membranes, electrodes, bipolar plates, and peripherals.

With the exception of the polymer membrane and platinum elements, which are the most critical components of a PEM fuel cell stack, in respect to durability, the other components of the PEM fuel cell system have a long lifetime since there are no moving parts. Because of this, in the present study it is considered that individual fuel cells will be replaced every 5 years.

Economic Assumptions

The following basic assumptions are made for the profitability assessment of the PEM fuel cell system:

1. lifetime: 15 years;
2. initial cost *per* unit of energy $I_{0,U}$: 1300 €/Kw$_E$ (PEM fuel cell system, materials, installation);
3. cost of hydrogen, CH_2: 10 €/GJ (fixed during project lifetime);
4. interest or discount rate, i: 7%;
5. annual fixed costs, AFC: 2.5% of I_0 (maintenance, insurance, interest charges of borrowed money, others);
6. cost of the replacement of the individual fuel cells at 5 and 10 years, CFCR: 25% of I_0;
7. PEM fuel cells operating time (hours a year), TO: 2000 h years^{-1} (at maximum capacity);
8. average annual PEM fuel cell efficiency, $\bar{\eta}$: 0.4
9. SV of the PEM fuel cell system, SV: 10% of I_0.

Considering a PEM fuel cell system of 10 kW$_E$ of power maximum capacity P_C, the initial investment is

$$I_0 = I_{0,U} * P_C = 1300 \frac{€}{kW_E} * 10\ kW_E = 13000\ €$$

The annual fixed costs are

$$A_{FC} = I_0 * 0.025 = 325\ €\quad years^{-1}$$

The annual electricity production AEP is

$$AEP = P_C * T_0 = 10\ kW_E * 2000\ h\ y^{-1} = 20000\ kW_E h\quad years^{-1}$$

Then, the annual fuel cost of hydrogen AFC$_{H2}$ during the first year can be calculated as

$$AFC_{H2} = \frac{C_H * AEP}{\bar{\eta}} = 1800\ €\,y^{-1}$$

By using Equation 8.31 and with the above considerations the NPC for the PEM fuel cell system is

$$NPC = \sum \frac{A_{TC}}{(1+i)^n} + I_0 = 35852.6 \text{ €}$$

In order to calculate the EUAC, it is necessary to estimate first the CRF, which can be calculated by Equation 8.16

$$CRF = \frac{i(1+i)^n}{(1+i)^n - 1} = \frac{0.07(1+0.07)^{15}}{(1+0.07)^{15} - 1} = 0.1098$$

Furthermore from Equation 8.35 the EUAC is

$$EUAC = NPV * CRF = 35852.6 \text{ €} * 0.1098 = 3936.6 \text{ €}$$

8.3.2 Profitability Assessment of a Compression Air Conditioning System

In order to determine the profitability of the proposed cogeneration systems in a PEM fuel cell coupled to an absorption air conditioning system, it is important first to analyze economically a compression air conditioning system in order to compare both technologies. The compression air conditioning system was chosen because it is well known that it is the most conventional air conditioning system, having been widely used for several decades.

Economic Assumptions

The following basic assumptions are made for the profitability assessment for the compression air conditioning system:

1. lifetime: 15 years;
2. initial cost *per* unit of cooling capacity, $I_{0,U}$: 200 €/KW$_C$ (compression air conditioning unit, materials, installation);
3. electricity unit cost CE: 0.14 €/kWh (average value of EU countries in 2007, Eurosat);
4. interest or discount rate, i: 7 %;
5. electricity inflation, i_E: 3 %;
6. annual fixed costs, AFC: 10 % of the initial cost I_0 (maintenance, insurance, interest charges on borrowed money, others);
7. annual cooling time required, ACTR: 1200 h years^{-1};
8. average yearly compressor coefficient of performance, COP = 2.5;
9. SV of the compression air conditioning: SV = 10 % of I_0.

Considering a compression air conditioning system of 10 kW of power capacity P_C, the initial investment is

$$I_0 = I_{0,U} * P_C = 200\frac{€}{kW} *10 \text{ kW} = 2000 \text{ €}$$

The annual fixed costs are

$$A_{FC} = I_0 *0.1 = 200 \text{ €} \quad \text{years}^{-1}$$

The annual cost of electricity A_E during the first year can be calculated as

$$A_E = \frac{P_C * A_{CTR} * C_E}{COP} = \frac{10 \text{ kW} *1200\frac{h}{y} *0.14\frac{€}{kWh}}{2}.5 = 672 \text{ € y}^{-1}$$

By using Equation 8.31 and with the considerations above the NPC for the compression air conditioning system is

$$NPC = \sum \frac{A_{TC}}{(1+i)^n} + I_0 = 11062.5 \text{ €}$$

In order to calculate the EUAC, it is necessary first to estimate the CRF, which can be calculated by Equation 8.16

$$CRF = \frac{i(1+i)^n}{(1+i)^n -1} = \frac{0.07(1+0.07)^{15}}{(1+0.07)^{15} -1} = 0.1098$$

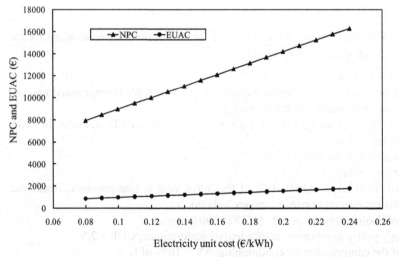

Figure 8.1 NPC and EUAC *versus* the electricity unit cost for a compression air conditioning system

Furthermore from Equation 8.35 the EUAC is

$$EUAC = NPV * CRF = 11062.5 \ € * 0.1098 = 1214.7 \ €$$

Figure 8.1 shows NPC and EUAC *versus* the electricity unit cost for a compression air conditioning system, keeping constant all the other parameters reported in the economic assumptions. The electricity unit cost varied from 0.8–0.24 €/kW, which are the limits of price of electricity in the EU. It can be observed that both the NPC and the EUAC economic parameters strongly depend on the electricity unit cost. The NPC varied from about 8000–0.08 €/kW h to 16000 € at 0.24 €/kW, while the EUAC varied from 870–1788 € for the same electricity unit costs.

8.3.3 Profitability Assessment of an Absorption Air Conditioning System

Economic Assumptions

The following basic assumptions were made for the profitability assessment of an absorption air conditioning system:

1. lifetime: 15 years;
2. initial cost *per* unit of cooling capacity $I_{0,U}$: 1000 €/KW$_C$ (absorption air conditioning unit, cooling tower, materials, installation);
3. gas unit cost CG: 10 €/GJ (average value of EU countries in 2007, Eurosat);
4. interest or discount rate, i: 7%;
5. gas inflation, i_G: 3 %;
6. annual fixed costs, AFC: 3 % of the initial investment I_0 (maintenance, insurance, interest charges of borrowed money, others);
7. annual cooling time required, ACTR: 1200 h years^{-1};
8. average yearly absorption coefficient of performance, COP = 0.65;
9. SV of the absorption unit and main equipment: SV = 10 % of I_0.

Considering an absorption air conditioning system of 10 kW of power capacity P_C, the initial investment is

$$I_0 = I_{0,U} * P_C = 1000 \frac{€}{kW} * 10 \ kW = 10000 \ €$$

The annual fixed costs are

$$A_{FC} = I_0 * 0.03 = 300 \ € \ y^{-1}$$

The annual cost of gas A_G during the first year can be calculated as

$$A_G = \frac{P_C * A_{CTR} * C_G}{COP} = \frac{10 \ kW * 1200 \frac{h}{y} * 3600 * 10 \frac{€}{GJ}}{0}.65 = 664.6 \ € \ y^{-1}$$

By using Equation 8.31 and with the above considerations, the NPC for the compression air conditioning system is

$$NPC = \sum \frac{A_{TC}}{(1+i)^n} + I_0 = 19602.8 \; €$$

To calculate the EUAC the CRF is first calculated by using Equation 8.16

$$CRF = \frac{i(1+i)^n}{(1+i)^n - 1} = \frac{0.07(1+0.07)^{15}}{(1+0.07)^{15} - 1} = 0.1098$$

Furthermore from Equation 8.35 the EUAC is

$$EUAC = NPV * CRF = 19602.8 \; € * 0.1098 = 2152.4 \; €$$

Figures 8.2–8.4 show the NPC and the EUAC *versus* the gas unit cost, the maintenance and operation costs, and the initial investment *per* unit of cooling capacity, respectively, for an absorption air conditioning system keeping constant all the other parameters reported in the economic assumptions.

In Figure 8.2 it can be seen that both economic parameters strongly depend on the gas unit cost. The NPC varied from 16709–26836 €, while the EUAC varied from 1834–2946 € for a gas unit cost from 6–20 €/GJ, respectively. Comparing the NPC and EUAC values with those obtained for the compression air conditioning system, reported in Figure 8.1, it can be observed that for the absorption system the economic parameters are considerably higher, which is mainly due to the higher initial costs.

Figure 8.3 shows the NPC and EUAC *versus* the maintenance and operation costs. It can be seen that both parameters are increased by the increment of the maintenance and operation costs; however, this increment is considerably lower

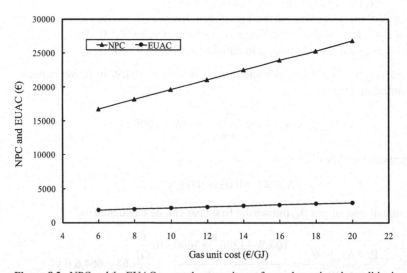

Figure 8.2 NPC and the EUAC *versus* the gas unit cost for an absorption air conditioning system

than that observed with the gas price. In this case, the NPC varied from 17781 to 22331 € at 100 €/years and 600 €/years, respectively, while with the gas unit cost the NPC varied from 16709 to 26836 €. Something similar can be observed with the EUAC values.

In Figure 8.4 it can be seen that the NPC and EUAC economic parameters depend strongly on the initial cost *per* unit of cooling capacity. The net present cost varied from 17129 to 22076 €, while the EUAC varied from 1880 to 2424 € for initial costs *per* unit of cooling capacity from 800 to 1200 €/kW$_C$, respectively.

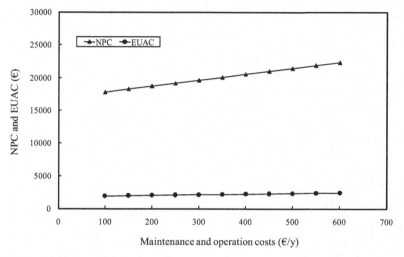

Figure 8.3 NPC and the EUAC *versus* the maintenance and operation costs for an absorption air conditioning system

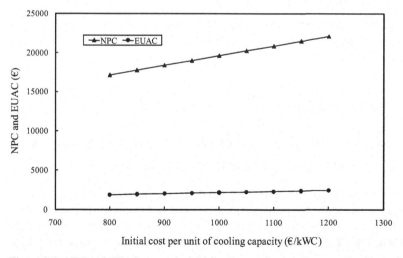

Figure 8.4 NPC and EUAC *versus* the initial cost *per* unit of cooling capacity for an absorption air conditioning system

8.3.4 Profitability Assessment for the PEMFC-CACS

In order to assess the profitability of the PEM fuel cell together with the conventional compression air conditioning system PEMFC-CACS the following assumptions were established.

Economic Assumptions

The following basic assumptions are made for the profitability assessment of the PEM fuel cell system:

1. lifetime: 15 years;
2. interest or discount rate, i: 7%;
3. electricity inflation, i_E: 3%;
4. electricity unit cost C_E: 0.14 €/kWh;
5. cost of hydrogen, CH_2: 10 €/GJ (fixed during project lifetime);
6. initial cost *per* unit of energy of the PEM fuel cell $I_{0,U}$: 1300 €/kW$_E$;
7. initial cost *per* unit of cooling capacity of the CACS, $I_{0,U}$: 200 €/kWc;
8. annual fixed costs of the PEM fuel cell, AFC: 2.5% of I_0;
9. cost of the replacement of the individual fuel cells at 5 and 10 years, CFCR: 25% of I_0;
10. annual fixed costs of the CACS, AFC: 10% of the initial cost I_0;
11. PEM fuel cells operating time (hours a year), TO: 2000 h years^{-1} (at maximum capacity);
12. ACTR: 1200 h years^{-1};
13. average annual PEM fuel cell efficiency, $\bar{\eta}$: 0.4;
14. average yearly compressor coefficient of performance, COP = 2.5;
15. SV of the PEMFC-CACS, SV: 10% of I_0.

Utilizing the calculated values for the PEM fuel cell and the compression air conditioning system in Sections 8.3.1 and 8.3.2, respectively and with the above economic assumptions, the NPC EUAC costs are the following.

The NPC

$$\text{NPC} = \sum \frac{A_{TC}}{(1+i)^n} + I_0 = 46915.1 \, €$$

and by using the capital recovery factor CRF from Equation 8.16, the EUAC is

$$\text{EUAC} = \text{NPV} * \text{CRF} = 46915.1 \, € * 0.1098 = 5151.3 \, €$$

Figure 8.5 shows the NPC and the EUAC *versus* the electricity unit cost for a compound system PEMFC-CACS. It can be observed that the NPC varied from almost 44000 € at an electricity unit cost of 0.08 €/kWh to about 52000 € at 0.24 €/kWh. The EUAC varied from 4800 to about 5700 € for the same electricity unit cost values.

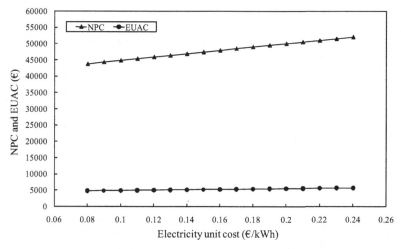

Figure 8.5 NPC and EUAC *versus* the electricity unit cost for the PEMFC-CACS

8.3.5 *Profitability Assessment for the PEMFC-AACS*

In order to assess the profitability of the cogeneration system consisting of a PEM fuel cell coupled to an absorption air conditioning system PEMFC-AACS the following assumptions were considered.

Economic Assumptions

The following basic assumptions are made for the profitability assessment of the PEM fuel cell system:

1. lifetime: 15 years;
2. interest or discount rate, i: 7 %;
3. cost of hydrogen, CH_2: 10 €/GJ (fixed during project lifetime);
4. initial cost *per* unit of energy of the PEM fuel cell $I_{0,U}$: 1300 €/kW$_E$;
5. initial cost *per* unit of cooling capacity of the AACS, $I_{0,U}$: 1000 €/kW$_C$;
6. annual fixed costs of the PEM fuel cell, AFC: 2.5 % of I_0;
7. cost of the replacement of the individual fuel cells at 5 and 10 years, CFCR: 25 % of I_0;
8. annual fixed costs of the AACS, AFC: 3 % of the initial cost I_0;
9. PEM fuel cells operating time (hours a year), TO: 2000 h years^{-1} (at maximum capacity);
10. annual cooling time required, ACTR: 1200 h;
11. average annual PEM fuel cell efficiency, $\bar{\eta}$: 0.4;
12. average yearly absorption coefficient of performance, COP = 0.65;
13. SV of the PEMFC-AACS, SV: 10 % of I_0.

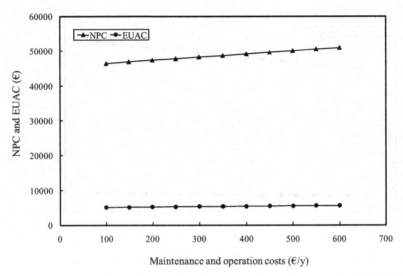

Figure 8.6 NPC and EUAC *versus* the maintenance and operation costs for the PEMFC-AACS

Utilizing the calculated values for the PEM fuel cell and the absorption air conditioning system in Sections 8.3.1 and 8.3.3, respectively, and considering that the cost of gas is equal to zero since all the energy needed for the operation of the absorption air conditioning system is provided by the heat released from the PEM fuel cell, and taking into account the above economic assumptions, the NPC and the EUAC are the following.

The NPC is

$$\mathrm{NPC} = \sum \frac{\mathrm{A_{TC}}}{(1+i)^n} + \mathrm{I}_0 = 48222.5 \text{ €}$$

and by using the CRF from Equation 8.16, the EUAC is

$$\mathrm{EUAC} = \mathrm{NPV} * \mathrm{CRF} = 46915.1 \text{ €} * 0.1098 = 5294.8 \text{ €}$$

Figure 8.6 shows the NPC and the EUAC *versus* the maintenance and operation costs for a cogeneration system consisting of a PEM fuel cell coupled to an absorption air conditioning system. It can be observed that the NPC varies from about 46400 € at a maintenance and operation cost of 100 €/years to almost 51000 € at 600 €/years. The EUAC varies from 5100 to about 5600 € for the same maintenance and operation costs.

Figure 8.7 shows NPC and EUAC *versus* the initial cost *per* unit of cooling capacity for the PEM fuel cell coupled to an absorption air conditioning system. It can be observed again that both the NPC and the EUAC economic parameters increase with the increment of the initial costs. The NPC varies from almost 46000 € at 800 €/kW$_C$ to about 51000 € at 1200 €/kW$_C$, while the EUAC varies from 5000 to almost 6000 € for the same initial costs.

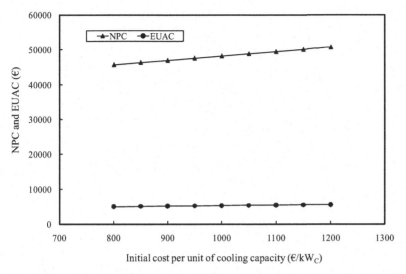

Figure 8.7 NPC and EUAC *versus* the initial cost *per* unit of cooling energy for the PEMFC-AACS

8.3.6 *Comparison of the Profitability Assessment of the PEMFC-AACS and the PEMFC-CACS*

In order to determine which of the analyzed systems is more profitable or under which conditions the proposed system is a better alternative from the economic point of view, in this section the PEMFC-AACS and the PEMFC-CACS are compared using the NPV, taking into consideration the economic assumption presented in the previous sections.

Figures 8.8 and 8.9 compare the NPC values for the PEMFC-AACS and the PEMFC-CACS *versus* the electricity unit price and maintenance and operation costs, and the electricity unit cost and the initial cost *per* unit of cooling capacity, respectively.

Figure 8.8 shows that the lowest NPC is obtained with the PEMFC-CACS at an electricity unit price of 0.08 €/kWh; however for higher values of the electricity price the NPC values increase more rapidly than those for the PEMFC-AACS with an increment of the maintenance and operation costs. At the reference value of the electricity unit price of 0.14 €/kWh the NPC for the PEMFC-CACS is 46915 €, while for the PEMFC-AACS they are 46400 and 46856 € at a maintenance and operation cost of 100 and 150 €/years, respectively. This means that for the electricity price of 0.14 €/kWh the PEMFC-AACS is more competitive if the maintenance and operation costs do not exceed the 150 € *per* year. The same way if the electricity price is 0.19 €/kW, the PEMFC-AACS is more competitive than the PEMFC-CACS whenever the maintenance and operation costs do not exceed 450 €

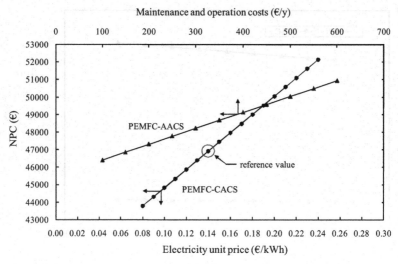

Figure 8.8 Comparison of the NPC values for PEMFC-AACS and PEMFC-CACS *versus* the electricity unit price and the maintenance and operation costs

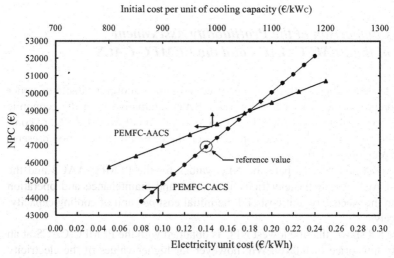

Figure 8.9 Comparison of the NPC values for PEMFC-AACS and PEMFC-CACS *versus* the electricity unit price and the initial cost *per* unit of cooling capacity

per year. Finally if the electricity price is 0.22 €/kW, then the PEMFC-AACS is always more competitive than the PEMFC-CACS.

Figure 8.9 shows the NPC for the PEMFC-AACS and PEMFC-CACS *versus* the electricity unit cost and the initial cost *per* unit of cooling capacity. It can be observed that for electricity unit costs up to 0.11 €/kWh, the PEMFC-CACS always has lower NPC than for the PEMFC-AACS; however at the reference value of 0.14 €/kWh the PEMFC-AACS is more competitive for initial cost *per* unit of cooling energy lower than 900 €. At the intersection point the PEMFC-AACS is a better alternative whenever the initial costs do not exceed the 1050 €, and for an electricity cost of 0.22 €/kWh the PEMFC-AACS is always better.

8.4 Conclusions

As can be observed from the profitability assessment of the cogeneration system (conformed by a PEM fuel cell and an absorption air conditioning system), actually there are some conditions in which the proposed system is more profitable than a PEM fuel cell and a compression air conditioning system working independently.

The NPC analysis showed that for an electricity unit price lower than 0.12 €/kW the cogeneration system proposed by using an absorption system is not profitable with the proposed parameters. However, if the electricity cost increases the cogeneration system it becomes a better alternative than the PEM fuel cell and the conventional compression system.

Even when in the profitability assessment for the reference case it was done taking an electricity price of 0.14 €/kWh, which is the average of the price in the EU, the electricity price varies for each country and in many of them the price is actually higher than the average value. This means that for those countries it is currently a better alternative to select the cogeneration system instead of the PEM fuel cell and the conventional compression air conditioning system.

It is important to point out that in the present analysis relevant factors related to the use of clean energies such as CO_2 bonus, incentives and preferential interest rates, others related to the size of the equipment and cost reduction due to higher production or market penetration, and other economic factors, were not taken into account. Furthermore, it is clear that if these parameters are properly estimated for a specific project, the proposed cogeneration system will be even more profitable in a higher number of countries.

In conclusion, the proposed cogeneration system consisting of a PEM fuel cell coupled to an absorption air conditioning system is actually, under certain circumstances, a good alternative for use were electricity and cooling are required and it will be even more profitable in the near future with the increase on the electricity prices.

References

Bar-On I, Kirchain R, Roth R (2002) Technical cost analysis for PEM fuel cells. J Power Sources 109:71–75
Chang HP, Chou CL, Chen YS, Hou TI, Weng BJ (2007) The design and cost analysis of a portable PEMFC UPS system. Int J Hydrogen Energy 32:316–322

Farbir F, Gómez T (1996) Efficiency and economics of proton exchange membrane (PEM) fuel cells. Int J Hydrogen Energy 21(10):891–901
Holland FA, Siqueiros J, Santoyo S, Heard CL, Santoyo ER (1999) Water purification using heat pumps. E & FN Spon, Taylor and Francis Group, London
Tsuchiya H, Kobayashi O (2004) Mass production cost of PEM fuel cell by learning curve. Int J Hydrogen Energy 29:985–990
Zoulias EI, Lymberopoulos N (2007) Techno-economic analysis of the integration of hydrogen energy technologies in renewable energy-based stand-alone power systems. Renew Energ 32:680–696

Index